Practical Approach to
Petrology

Second Edition

Practical Approach to
Petrology

Second Edition

Rabindra Nath Hota

Professor
PG Department of Geology
Utkal University, Vani Vihar
Bhubaneswar
Odisha

CBSPD

CBS Publishers & Distributors Pvt Ltd

New Delhi • Bengaluru • Chennai • Kochi • Kolkata • Lucknow • Mumbai
Hyderabad • Jharkhand • Nagpur • Patna • Pune • Uttarakhand

Practical Approach to
Petrology
Second Edition

ISBN: 978-93-86217-67-7

Copyright © Author and Publisher

Second Edition 2017
  Reprint: 2023
First Edition 2011
  Reprint: 2012

Published by Satish Kumar Jain and produced by Varun Jain for

CBS Publishers & Distributors Pvt Ltd

4819/XI Prahlad Street, 24 Ansari Road, Daryaganj, New Delhi 110 002, India
Ph: 011-23289259, 23266861   Website: www.cbspd.com
      e-mail: delhi@cbspd.com

Corporate Office: 204 FIE, Industrial Area, Patparganj, Delhi 110 092
Ph: 011-4934 4934   Fax: 011-4934 4935   e-mail: publishing@cbspd.com; publicity@cbspd.com

Branches

- **Bengaluru:** Seema House 2975, 17th Cross, K.R. Road, Banasankari 2nd Stage, Bengaluru 560 070, Karnataka, India
  Ph: +91-80-26771678/79   Fax: +91-80-26771680 e-mail: bangalore@cbspd.com
- **Chennai:** 7, Subbaraya Street, Shenoy Nagar, Chennai 600 030, Tamil Nadu, India
  Ph: +91-44-26680620, 26681266 Fax: +91-44-42032115 e-mail: chennai@cbspd.com
- **Kochi:** 42/1325, 1326, Power House Road, Opp KSEB, Power House, Ernakulam 682 018, Kerala, India
  Ph: +91-484-4059061-65   Fax: +91-484-4059065 e-mail: kochi@cbspd.com
- **Kolkata:** 147, Hind Ceramics Compound, 1st Floor, Nilgunj Road, Belghoria, Kolkata-700056, West Bengal, India
  Ph: 033-25633055, 033-25633056   e-mail: kolkata@cbspd.com
- **Lucknow:** Basement, Khushnuma Complex, 7-Meerabai Marg (Behind Jawahar Bhawan) Lucknow 226001, India
  Ph: 0522-4000032   e-mail: tiwari.lucknow@cbspd.com
- **Mumbai:** PWD Shed. Gala no. 25/26, Ramchandra Bhatt Marg, Next to JJ Hospital Gate no. 2, Opp. Union Bank of India Noorbaug Mumbai-400009, Maharashtra, India
  Ph: 022-66661880/89   e-mail: mumbai@cbspd.com

Representatives

- **Hyderabad** 0-9885175004 • **Jharkhand** 0-9811541605 • **Nagpur** 0-9421945513
- **Patna** 0-9334159340 • **Pune** 0-9923910676 • **Uttarakhand** 0-9716462459

Printed at Rashtriya Printers, Dilshad Garden, Delhi, India

Foreword

Very few authors venture to write books on practical aspects of geology because of vast scope of the subject. This problem becomes more acute while dealing with exhaustive subject like petrology that deals with igneous, sedimentary and metamorphic petrology including that of petrochemistry. The long felt need of a book encompassing all the branches of petrology has taken the shape in form of *Practical Approach to Petrology* written by Dr RN Hota.

In this endeavour, Dr Hota has dealt with all the aspects of petrology including that of petrochemistry. Some of the problems enumerated in this book has been worked out, whereas some problems are given at the end for solution by the students. The practical aspects of sedimentary petrology have been given in an exhaustive manner and dealt in this book covers major part of the theoretical aspects of sedimentary petrology.

The book by Dr RN Hota is a welcome addition to the library on petrology. I strongly recommend and endorse this publication to our geosciences fraternity.

HK Sahoo
Professor
PG Department of Geology
Utkal University, Vani Vihar
Bhubaneswar
Odisha

Preface to the Second Edition

The book *Practical Approach to Petrology* published in January 2011 got reprinted in 2012, which signifies the awesome response from the readers, particularly the students for whom it has been written. I thank all the readers for their patronage. The stray typographical and other errors of the first edition have been eradicated in this edition. In addition, many of the paragraphs have been rewritten and rearranged to maintain homogeneity of the chapters and also for better readability. Ichno fossils, palaeohydrology and problems in grain size analysis, tilt correction and palaeocurrent have been added to make the book complete.

I sincerely appeal the readers (teachers and students) to suggest for betterment of this book in future. They are requested to send their feedbacks and suggestions to me through e-mail (rnhota@yahoo.com).

Rabindra Nath Hota

Preface to the Second Edition

Preface to the First Edition

The thought of writing a textbook in geology crept into my mind in 1987, when I started my teaching career as a lecturer in the undergraduate section of the Department of Geology, Utkal University located at Ravenshaw College, Cuttack. I could realise the problems faced by students who join the degree classes directly to make geology as their career. Further, shortage of teachers and supporting staff makes the problem more acute. After being free from conventional research for PhD and Post-Doctoral research works, I initiated the writing work. Like other science subjects, both theory and practical are equally important in geology in addition to the field study. A few textbooks and a number of reference books on the theory part of the subject are available in Indian market. Realising the difficulties faced by the students in practical classes, I started writing textbook on practical aspects of different branches in geology keeping the revised curriculum of the University Grants Commission in mind.

All the three branches of petrology are fundamental branches of geology. The igneous petrology part of this book deals with the study of texture, structure, petrography including calculation of norm, modal analysis and graphic plot of petrochemistry data. A good number of problems have been added to make the students conversant with norm calculation. The metamorphic petrology part includes texture, structure, petrography including construction of ACF, AKF and AFM diagrams. The sedimentary geology branch embodies texture, structure, petrographic study of rocks as well as grain size, palaeocurrent and heavy mineral analyses.

In the process of writing, a number of books written by different authors have been followed including incorporation of handouts provided to the students in practical classes. Any part of the text, method, statement and illustrations incorporated in the writing of this book is acknowledged. I am indebted and thankful to the authors and publishers of those referred materials. I express my deep sense of gratitude to Prof. Wataru Maejima of Osaka City University, Japan, who taught me the methods of drawing/redrawing of figures, as a result of which the book could be illustrative with a lot of figures. I am thankful to Prof. HK Sahoo of PG Department of Geology, Utkal University, who read major part of the manuscript, gave valuable suggestions to make the book complete in all respect and kindly agreed to write the Foreword to this book. However, I accept the responsibility of any mistake, which might have been there inadvertently. In spite of critical scrutiny, some typographical errors might have been left out. I shall be very much thankful, if, lapses of any kind are communicated to me (rnhota@yahoo.com).

I express my gratitude to Prof. (Mrs) M Das, Dr PP Singh, Dr BK Ratha, Mr MR Mohapatra and Mr G Das of the PG Department of Geology, Utkal University for their encouragement and help during preparation of the manuscript.

I am thankful to Mr YN Arjuna, Senior Director, Publishing, Editorial and Publicity and M/s CBS Publishers & Distributors, New Delhi for their keen interest in publishing the book.

Last but not the least, I thank my wife Anjali and children Swayam Prakash and Soumya Ranjan for their inspiration and cooperation during preparation of the manuscript.

Rabindra Nath Hota

Contents

Foreword by HK Sahoo — v

Preface to the Second Edition — vii

Preface to the First Edition — ix

PART I: IGNEOUS PETROLOGY

1. Structure of Igneous Rocks — 3

    1.1 Blocky Structure 3

    1.2 Ropy Lava 4

    1.3 Pillow Structure 4

    1.4 Vesicular Structure 4

    1.5 Amygdaloidal Structure 4

    1.6 Flow Structure 4

    1.7 Jointing and Platy Structure 4

    1.8 Columnar and Prismatic Structure 5

2. Texture of Igneous Rocks — 6

    2.1 Crystallinity 6

    2.2 Granularity 6

    2.3 Shape of Crystals 6

    2.4 Mutual Relationship of Crystals 7

        2.4.1 Equigranular Texture 7

        2.4.2 Inequigranular Texture 7

    2.5 Microstructure 8

        2.5.1 Reaction Structure 8

        2.5.2 Xenolithic Structure 8

        2.5.3 Orbicular Structure 9

        2.5.4 Spherulitic Structure 9

        2.5.5 Fracture Form 9

3. Petrography of Igneous Rocks — 10

4. **Calculation of Norm**   **17**
    4.1   Rules for Calculation of Norm   17
    4.2   Normative Classification of Igneous Rocks   22
    4.3   Calculation of Niggli Values and Other Parameters   23
    4.4   Significance of Niggli Values   24
    4.5   Norm Calculation of an Oversaturated Rock   25
    4.6   Norm Calculation of an Undersaturated Rock   29
    4.7   Problems for Norm Calculation   34
    4.8   Answers   36

5. **Modal Analysis**   **40**
    5.1   Modal Analysis of an Igneous Rock   41
    5.2   Modal Analysis of a Sedimentary Rock   42

6. **Graphic Plot of Petrochemistry Data**   **44**
    6.1   Variation Diagram   44
    6.2   Triangular Diagram   46
    6.3   Discrimination Diagram   47

PART II: METAMORPHIC PETROLOGY

7. **Fabric and Petrography of Metamorphic Rocks**   **53**
    7.1   Shape of Metamorphic Minerals   53
    7.2   Metamorphic Texture   53
          7.2.1   Lepidoblastic Texture   54
          7.2.2   Nematoblastic Texture   54
          7.2.3   Crystalloblastic Texture   54
          7.2.4   Granoblastic Texture   54
          7.2.5   Porphyroblastic Texture   54
          7.2.6   Poikiloblastic Texture   54
          7.2.7   Cataclastic Texture   54
          7.2.8   Hornfelsic Texture   54
    7.3   Metamorphic Structure   54
          7.3.1   Granulose Structure   54
          7.3.2   Schistose Structure   55
          7.3.3   Gneissose Structure   55
          7.3.4   Maculose Structure   55

    7.3.5 Cataclastic Structure 55
    7.3.6 Phyllitic Structure 55
    7.3.7 Slaty Structure 55
  7.4 Petrography of Metamorphic Rocks 55

8. **Graphic Construction of ACF, AKF and AFM Diagrams**   **59**
  8.1 ACF Diagram 59
  8.2 AKF Diagram 61
  8.3 AFM Diagram 62
  8.4 Use of ACF, AKF and AFM Diagrams 64

PART III: SEDIMENTARY PETROLOGY

9. **Texture of Sedimentary Rocks**   **67**
  9.1 Grain Size 67
  9.2 Sorting 68
  9.3 Shape 68
  9.4 Roundness 70
  9.5 Fabric 72
  9.6 Packing 72
  9.7 Clastic Texture 73
  9.8 Nonclastic Texture 73
  9.9 Organic Texture 73
  9.10 Surface Texture 73
  9.11 Matrix and Cement 74
  9.12 Microscopic Study of Diagenetic Features 74

10. **Structure of Sedimentary Rocks**   **76**
  10.1 Mechanical or Primary Structures 76
    10.1.1 Bedding/Lamination 76
    10.1.2 Ripple Mark 77
    10.1.3 Cross Lamination 78
    10.1.4 Cross Bedding 78
    10.1.5 Flaser, Wavy and Lenticular Beddings 79
    10.1.6 Swash Cross-stratification 79
    10.1.7 Hummocky Cross-stratification 80
    10.1.8 Herringbone Cross-stratification 80

10.1.9   Graded Bedding   81
10.1.10 Sole Marks   81
10.1.11 Mud Crack Cast   82
10.1.12 Rib and Furrow Structure   83
10.1.13 Rain Prints   83
10.1.14 Armored Mud Balls   83
10.2   Chemical or Secondary Structures   83
10.2.1   Nodule   83
10.2.2   Spherulite   83
10.2.3   Rosette   84
10.2.4   Concretions   84
10.2.5   Geodes   84
10.2.6   Septaria   84
10.2.7   Cone-in-cone   84
10.2.8   Stylolite   85
10.2.9   Corrosion Surface   85
10.2.10 Vug   85
10.2.11 Oolicast   85
10.2.12 Crystal Mould   86
10.3   Organic or Biogenic Structures   86
10.3.1   Petrification   86
10.3.2   Tracks and Trails   86
10.3.3   Burrows and Borings   87
10.3.4   Faecal Pellets   87
10.3.5   Coprolites   87
10.3.6   Stromatolites   87
10.4   Trace Fossils or Ichnofossils   88

11.   Classification of Sedimentary Rocks   **100**
11.1   Pettijohn's Classification   100
11.2   Friedman and Sander's Classification   100
11.3   Classification of Conglomerates and Breccias   101
11.4   Classification of Sandstones (Dott, 1964; Folk, 1968)   102
11.5   Classification of Mud Rocks   102
11.6   Classifications of Carbonate Rocks (Limestone)   103
11.7   Classification of Sedimentary Aggregates   104
11.8   Petrography of Sedimentary Rocks   104

12. Grain Size Analysis   108

12.1   Method of Grain Size Analysis   108

    12.1.1   Disaggregation of Grains   111

    12.1.2   Sieving   111

    12.1.3   Pipette Analysis   111

    12.1.4   Analysis of Grain Size Data   112

12.2   Significance of Grain Size Study   117

12.3   Problems for Solution   119

12.4   Answers   120

13. Palaeocurrent and Palaeohydrological Analysis   124

13.1   Scalar Properties of Sediments and Palaeocurrent   124

13.2   Vector Properties of Sediments and Palaeocurrent   126

13.3   Tilt Correction   126

    13.3.1   Tilt Correction for Planar Structures   126

    13.3.2   Tilt Correction for Linear Structures   127

13.4   Tilt Correction Problems   127

    13.4.1   Problems   127

    13.4.2   Answers   128

    13.4.3   Problems   128

    13.4.4   Answers   128

13.5   Computation of Mean Palaeocurrent Direction and Other Statistics   128

    13.5.1   Worked Out Examples   130

13.6   Graphic Representation of Palaeocurrent Data   133

13.7   Statistical Significance of Resultant Palaeocurrent Direction   135

13.8   Comparison of Palaeocurrent Populations of Two Formations   138

13.9   Palaeocurrent Problems for Solution   140

13.10   Palaeohydrological Analysis   142

    13.10.1 Problems for Solution   144

    13.10.2 Answers   144

13.11   Significance of Palaeocurrent Analysis   145

14. Heavy Mineral Analysis   146

14.1   Methods of Heavy Mineral Separation   146

    14.1.1   Hand Picking Method   147

    14.1.2   Magnetic Method   147

    14.1.3   Electrical Method   148

    14.1.4   Heavy Liquid Method   148

14.2 Washing and Mounting of Heavy Minerals 148
14.3 Identification of Heavy Minerals 149
14.4 Counting of Heavy Minerals 149
14.5 Graphical Presentation of Heavy Mineral Frequency 160
14.6 Significance of Heavy Mineral Study 161

Bibliography 163

Index 165

Part I

Igneous Petrology

1. Structure of Igneous Rocks

2. Texture of Igneous Rocks

3. Petrography of Igneous Rocks

4. Calculation of Norm

5. Modal Analysis

6. Graphic Plot of Petrochemistry Data

1

Structure of Igneous Rocks

Magma is a viscous mass of molten silicates formed at high temperature beneath the earth's crust. It also contains different types of gases and water. The rocks, which are formed by cooling of magma, are called igneous rocks. Depending on the depth of cooling, the igneous rock is said to be plutonic, if the depth is appreciably high or hypabyssal, if the depth is low. When the magma is erupted to the surface it is termed *lava*, which on cooling forms volcanic rock. Depending on the composition there are two types of primary magmas, granitic and basaltic. The granitic magma is saturated with silica (SiO_2) and in normal cooling condition yields granite. The basaltic magma, on the other hand, is deficient in silica and gives rise to basalt. Other types of igneous rocks are formed from these magmas by the processes of differentiation and assimilation.

The chemical composition of the magma determines the minerals to be formed while the physical conditions prevailing at the time of consolidation control the degree of crystallization (crystallinity), absolute size of the crystals/grains (granularity), shape and mutual relationship among the crystals. These factors are collectively known as *texture* of the igneous rocks. Juxtaposition of two or more kinds of textures is known as *microstructure*. Reaction, orbicular and spherulitic structures belong to this category. The structures in the true sense, refer to large-scale features like blocky and pillow structures, flow banding, jointing, etc. Thus, an igneous rock is characterised by its composition (constituent minerals), texture and structure. The structure and texture, which are used for identification of the rocks in hand specimen (megascopic identification) and in thin section (microscopic identification), are described in this and subsequent chapters.

The structure refers to large-scale features produced on the surface of igneous rocks particularly in the volcanic type, which is erupted to the surface. Most of them are better visualized in the field. Some important structures are described below.

1.1 BLOCKY STRUCTURE

The term is applied when the surface of lava flow is covered with a mass of rough, jagged and angular blocks of all dimensions.

1.2 ROPY LAVA

Highly mobile lavas solidify with glazed, smooth or wrinkled surfaces. The surfaces are often marked with low domes of a few meters in diameter and characteristic radial cracks.

1.3 PILLOW STRUCTURE

This refers to the appearance of pillow shaped small masses on the surface of soda-rich basaltic lava (spilites). The pillow has a vesicular crust with occasional glassy skin. The masses are generally elongated like bolsters and exhibit parallelism of their long axes.

1.4 VESICULAR STRUCTURE

Most of the lavas contain appreciable amount of gases, which escape soon after eruption of the lava to the surface. The escape of the gases produces cavities or vesicles, which are spherical, cylindrical, elliptical or irregular in shape. When the vesicles are enormously high, the term *scoria* is applied. *Pumice* is produced when the frequency of vesicles is still much higher.

1.5 AMYGDALOIDAL STRUCTURE

When the vesicles are filled with secondary minerals like zeolites, calcite or different forms of silica, the structure is known as amygdaloidal structure.

1.6 FLOW STRUCTURE

No lava is completely homogenous during and immediately after eruption to the surface. Patches and layers in it differ in composition, which are drawn out into parallel lenticles or bands of different colour and texture as the lava flows down resulting in the formation of flow structure. Acid lavas like rhyolite and trachyte generally exhibit this type of structure. Banded structure is also seen in some plutonic rocks due to alternation of layers differing in mineral composition and/or texture. The differentiation and/or assimilation processes generally produce this type of structure, which make the homogeneous magma heterogeneous.

1.7 JOINTING AND PLATY STRUCTURE

Joints are divisional planes found in all types of igneous rocks. Three sets of joints, one horizontal and two vertical (Fig. 1.1) are commonly seen in granites that result in cuboidal blocks. Close spacing of the horizontal joint planes result in the formation of sheet or platy structure. Occasionally, the joint planes are curved or undulating. In such cases the rocks are divided into spherical or ellipsoidal blocks, at times, resembling the pillow structure. Jointing is commonly caused by tensile stress consequent upon contraction due to cooling or by tectonic causes producing tensional, compressional or torsional stresses.

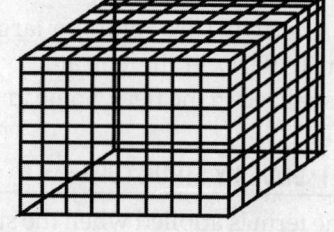

Fig. 1.1: Joint and platy structure

1.8 COLUMNAR AND PRISMATIC STRUCTURE

With uniform cooling, centers of contraction at equal spaced intervals develop in igneous rocks (Fig. 1.2). The lines joining these centers are the directions of greatest tensile stress. When the tensile stress exceeds the rigidity of the rock, cracks appear perpendicular to the directions of greatest tensile stress. The cracks extend upwards and downwards resulting in the formation of columnar prismatic forms. The columns may be four to seven-sided often intersected by cross-joints. A six-sided columnar structure is shown in Fig. 1.3.

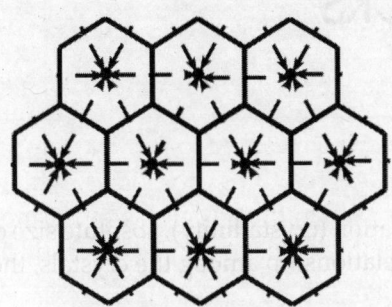

Fig. 1.2: Centers of contraction

Fig. 1.3: Columnar structure

2

Texture of
Igneous Rocks

Texture of igneous rock refers to degree of crystallization (crystallinity), absolute size of the crystals/grains (granularity), shape and mutual relationship among the crystals, the last two are collectively known as fabric of the rock.

2.1 CRYSTALLINITY

Crystallinity is the measure of degree of crystallization. A rock is said to be holocrystalline when it is composed entirely of crystals and holohyaline when it is composed entirely of glass. The terms mero-, hypo- and hemi-crystalline are used when the rock is composed of both crystals and glass. The rate of cooling and viscosity of magma are the chief factors which determine the crystallinity. Rapid cooling and high viscosity favour the formation of glass, while slow cooling and low viscosity lead to the formation of crystals. The holocrystalline texture is the characteristic of deep-seated plutonic rocks. The merocrystalline texture, on the other hand, is seen in hypabyssal rocks, which cool at a shallow depth. Holohyaline texture is rare in occurrence and is seen in marginal parts of lava flow, which undergo quick chilling.

2.2 GRANULARITY

The grain size of igneous rocks show wide variation ranging from submicroscopic as in case of microlites to very large, measurable in meters as in case of some pegmatites. The rock is said to be phanerocrystalline or phaneric when the crystals are big enough to be visible in naked eye. The terms coarse-, medium- and fine-grained are used for grain sizes larger than 5 mm, 5–1 mm and less than 1 mm respectively. The term aphanitic is used when the grains cannot be visualized in naked eye. Aphanitic rocks may be *microcrystalline* when the grains can be distinguished under microscope or *cryptocrystalline* when the grains cannot be distinguished under microscope.

2.3 SHAPE OF CRYSTALS

Depending on the degree of development of crystal faces, the terms euhedral, anhedral and subhedral are used. Euhedral refers to the development of well-developed crystal faces while the term anhedral is used for complete lack of faces. The term subhedral is used for intermediate

6

stage. Depending on the relative dimensions, the crystals may be described as equidimensional, tabular or prismatic. Equidimensional crystals show uniform growth in all directions as in case of garnet, leucite, augite, etc. Tabular or bladed crystals are better developed in two dimensions than the third as in case of kyanite, micas, etc. Crystals better developed in one direction in comparison to other two are termed prismatic as in case of hornblende, sillimanite, etc.

2.4 MUTUAL RELATIONSHIP OF CRYSTALS

A number of textural terms are used to describe the mutual relationship among crystals. These are as follows.

2.4.1 Equigranular Texture

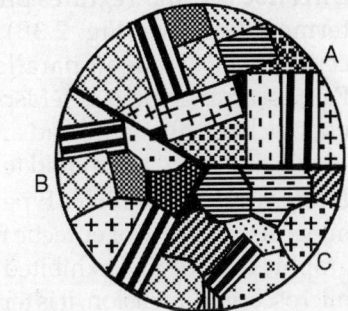

In case of equigranular texture, the constituent minerals are approximately of the same size and the rock is evenly granular. When most of the grains are euhedral, as in case of lamprophyres and carbonatites, the term panidiomorphic (Fig. 2.1A) is used. If majority of the grains are subhedral as in case of granites and syenites, the texture is said to be hypidiomorphic (Fig. 2.1B). The term allotriomorphic is used to describe a rock having larger proportion of anhedral grains (Fig. 2.1C).

Fig. 2.1: Igneous textures: A. Panidio-morphic, B. Hypidiomorphic, C. Allotrio-morphic

2.4.2 Inequigranular Texture

In this case the constituent mineral grains differ in size. Porphyritic and poikilitic are two important textures belonging to this category.

i. *Porphyritic texture*: In case of porphyritic texture, large crystals (phenocrysts) are surrounded by a groundmass of smaller crystals (Fig. 2.2A). The texture is termed megaporphyritic when it is seen with naked eye and microporphyritic when it is distinguishable under microscope. The porphyritic texture is termed as felsophyric and vitrophyric depending on the nature of groundmass, i.e. whether the groundmass is cryptocrystalline or glassy respectively.

ii. *Poikilitic texture*: The texture is said to be poikilitic when smaller crystals are enclosed within larger crystals without any preferred orientation (Fig. 2.2B). The inclusions must be numerous enough to produce a distinctive pattern. Common and minute inclusions of apatite and zircon, which occur within rock forming mineral like quartz are not considered under this category. The ophitic texture is a special type of poikilitic texture in which numerous plagioclase crystals are found to be enclosed within augite (Fig. 2.2C). This type of texture is common in dolerite. In some dolerites, the plagioclase crystals are large enough so that their enclosure is partial. In such case, the texture is termed subophitic.

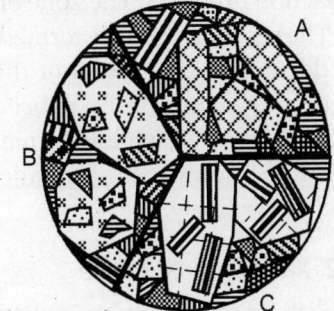

Fig. 2.2: Igneous textures: A. Porhyritic, B. Poikilitic, C. Ophitic

iii. *Intergranular and intersertal textures*: In many types of basalts, the plagioclase laths are so arranged that triangular or polygonal spaces are left within the crystals. When these interspaces are filled with granular augite, olivine or iron oxide the resulting texture is known as intergranular (Fig. 2.3A). Conversely, if the interspaces are filled with glassy, cryptocrystalline or fine-grained chlorite or serpentine, the texture is known as intersertal.

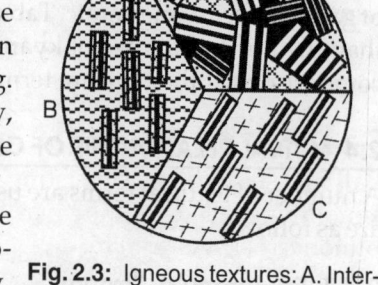

iv. *Directive texture*: Textures produced in flowing lavas are termed directive (Fig. 2.3B). The early formed micro-crystals are arranged parallel to the direction of flow. Parallel arrangement of feldspar laths in feldspathic lavas

Fig. 2.3: Igneous textures: A. Inter-granular, B. Directive, C. Perthite

like trachyte, phonolite and andesite is termed trachytic, whereas similar arrangement in syenite is termed trachytoid texture.

v. *Intergrowth texture*: This type of texture results due to simultaneous crystallization of two minerals as in case of eutectic melts. The intergrowth of orthoclase and quartz is known as graphic texture well exhibited by graphic granite. When the graphic texture is reduced to microscopic dimension, it is termed micrographic and the rock is known as micropegmatite. Another well-known intergrowth texture is known as perthite (and microperthite) in which albite crystals are enclosed within potash feldspars like orthoclase or microcline (Fig. 2.3C). In the reverse case, when albite crystal encloses orthoclase or microcline, the term antiperthite is used. The term myrmekite is used for intergrowth of quartz and plagioclase feldspar, in which the quartz occurs as worm-like rods within plagioclase feldspar.

2.5 MICROSTRUCTURE

Reaction, xenolithic, orbicular and spherulitic structures belong to this category.

2.5.1 Reaction Structure

In many instances reaction takes place between early formed minerals and magma. In the process, the reacting mineral may completely disappear. In some cases, the reaction is incomplete and the corroded crystal is found to be surrounded by reaction products. The zone of reaction is known as reaction rim. The reaction rims are termed corona (Fig. 2.4A) or kelyptic boarder depending whether they are produced by primary magmatic reaction or by subsequent metamorphic process. Reaction rims are generally formed around olivine, hypersthene and garnet leading to the formation of pyroxene, amphibole and spinel respectively.

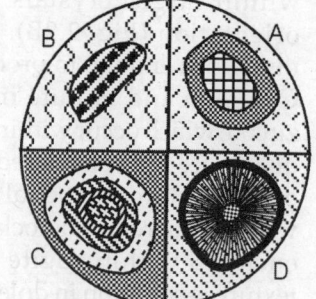

2.5.2 Xenolithic Structure

Xenolith or enclave is an inclusion of foreign rock fragment within an igneous rock (Fig. 2.4B). It may be genetically related to the enclosing igneous rock, i.e. formed in the early stage of

Fig. 2.4: Microstructures: A. Reaction, B. Xenolithic, C. Orbicular, D. Spherulitic

crystallization. In such case the xenolith is said to be cognate. However, in many instances the xenolith is a fragment of any pre-existing country rock, which has been accidentally incorporated into the invading magma or lava flow. In such case the xenolith is termed accidental. The size of xenolith varies from submicroscopic to several meters in diameter.

2.5.3 Orbicular Structure

Ball like segregation consisting of concentric shells of different minerals with or without a xenolithic nucleus is known as orbicular structure (Fig. 2.4C). These are generally found in plutonic rock like granite.

2.5.4 Spherulitic Structure

Spherulitic structure consists of fibers radiating out from a common center (Fig. 2.4D). It may be completely spherical or a part of a sphere. The size varies from microscopic to large masses of tens of centimeters in diameter. Spherulites generally occur in acid volcanic or hypabyssal rocks. They have also been reported from basic lavas and intrusions where they are called variolites and the rock containing them, variolite.

2.5.5 Fracture Form

Fracture form develops in volcanic glass due to contraction resulting from quick chilling. The rock showing fracture form appears like an aggregate of nodular masses consisting of concentric onion-like shells. Another type of fracture form consisting of irregularly radiating fissures develops in neighbouring minerals due to release of stress associated with increase of volume of a mineral undergoing alteration. This type of fracture is well marked in feldspar associated with olivine altered to serpentine.

3

Petrography of
Igneous Rocks

Petrography refers to the systematic description of rocks in hand specimen and in thin section. In case of igneous rocks, the name of the rock not only depends on the mineralogical constituent but also on the grain size. The mineral composition depends on the composition of the magma, whereas the grain size depends on the rate of cooling, which in turn is a function of depth. Thus, some rocks of similar mineralogical composition showing different grain sizes are given different names. For example, the common rock granite contains quartz and alkali feldspars as essential minerals and is coarse-grained. Rocks of same mineral composition and fine- to very fine-grained are termed felsite and rhyolite respectively. Similarly, dolerite and basalt are the hypabyssal (medium-grained) and volcanic (fine-grained) equivalents of gabbro/norite respectively.

Grain size and essential mineral constituents of some igneous rocks are given in Table 3.1 and detailed petrography of common igneous rocks is presented in Table 3.2.

TABLE 3.1: Different types of igneous rocks with essential minerals and grain size		
Rock	Essential minerals	Grain size
Acid-charnockite	Hypersthene + quartz + orthoclase	Coarse-grained
Adamellite	Quartz + alkali feldspar + plagioclase	Coarse-grained
Aegirinite	Aegirine	Coarse-grained
Allivalite	Anorthite + olivine	Coarse-grained
Alnoite	Biotite + feldspathoids	Medium-grained
Andesite	Plagioclase (dominant) + potash feldspar	Fine-grained
Ankaramite	Augite + olivine – basalt	Fine-grained
Anorthosite	Plagioclase (labradorite, bytwonite)	Coarse-grained
Aplite	Potash feldspar + plagioclase + quartz	Fine-grained
Basalt	Labradorite + pyroxene	Fine-grained
Basanite	Olivine and feldspathoid – basalt	Fine-grained
Basic-charnockite	Hypersthene + plagioclase + diopside	Coarse-grained
Borolanite	Pseudo-leucite + orthoclase + melanite + aegirine + biotite	Fine-grained
Bronzitite	Bronzite	Coarse-grained
Camptonite	Barkevikite/augite + plagioclase	Coarse-grained
Carbonatite	Calcite + dolomite	Coarse-grained

(Contd...)

TABLE 3.1: Different types of igneous rocks with essential minerals and grain size (*Contd...*)

Rock	Essential minerals	Grain size
Charnockite	Quartz + microcline + oligoclase + hypersthene	Coarse-grained
Cortlandite	Olivine + amphibole	Coarse-grained
Dacite	Quartz + potash feldspar < plagioclase	Fine-grained
Diabase	Labradorite + augite, often altered to amphibole	Coarse-grained
Diallagite	Augite +diopside + pigeonite	Coarse-grained
Diorite	Plagioclase (dominant) + potash feldspar	Coarse-grained
Ditrolite	Nepheline + perthite + sodalite + aegirine + Na-amphibole	Medium-grained
Dolerite	Augite + labradorite	Medium-grained
Dunite	Olivine	Coarse-grained
Eclogite	Garnet (pyrope, almandine) + chrome-diopside (omphasite)	Coarse-grained
Enderbite	Charnockitic-granodiorite	Coarse-grained
Essexite	Labradorite + orthoclase + augite + hornblende + biotite	Coarse-grained
Eucrite	Anortite + pyroxene	Medium-grained
Felsite	Potash and plagioclase feldspars	Fine-grained
Fergusite	Pseudo-leucite + aegirine	Medium-grained
Forsteritite	Forsterite	Coarse-grained
Foyaite	Nepheline + perthite + aegirine	Coarse-grained
Fyalite-syenite	Fyalite + plagioclase	Coarse-grained
Gabro	Labradorite + clino-pyroxene (augite, diopside, pigeonite)	Coarse-grained
Granite	Quartz + potash feldspar > plagioclase	Coarse-grained
Granodiorite	Quartz + potash feldspar < plagioclase	Coarse-grained
Granophyre	Quartz + potash feldspar > plagioclase	Medium-grained
Harzburgite	Olivine + ortho-pyroxene	Coarse-grained
Hornblendite	Hornblende	Coarse-grained
Hortonolitite	Hortonolite	Coarse-grained
Hypersthenite	Hypersthene	Coarse-grained
Ijolite	Nepheline + Na-pyroxene	Coarse-grained
Intermediate-charnockite	Hypersthene + potash feldspar	Coarse-grained
Kersantite	Biotite + plagioclase	Coarse-grained
Kimberlite	Phlogopite + bronzite + chrome-diopside	Coarse-grained
Larvikite	Anorthoclase (intergrowth of orthoclase and oligoclase)	Coarse-grained
Leucite basanite	Alkali basalt with leucite + olivine + alkali feldspar	Fine-grained
Leucite tephrite	Alkali basalt with leucite + alkali feldspar	Fine-grained
Leucite-syenite	Leucite + alkali feldspar	Coarse-grained
Leucitite	Alkali basalt with leucite	Fine-grained
Leucitophyre	Trachytes with leucite	Fine-grained
Lherzolite	Olivine + pyroxene	Coarse-grained
Litchfieldite	Albite + microcline + nepheline + perthite + aegirine + sodalite/cancrinite	Fine-grained
Malignite	Nepheline + orthoclase + aegirine	Fine-grained
Mariupolite	Albite + nepheline + aegirine/Na-amphibole	Fine-grained
Micro-adamellite	Quartz + potash feldspar <2/3 + plagioclase <2/3	Medium-grained
Micro-diorite	Plagioclase (dominant) + potash feldspar	Medium-grained
Micro-granodiorite	Quartz + potash feldspar <1/3 + plagioclase >2/3	Medium-grained
Micro-monzonite	Potash feldspar + plagioclase	Medium-grained

(*Contd...*)

TABLE 3.1: Different types of igneous rocks with essential minerals and grain size (*Contd...*)

Rock	Essential minerals	Grain size
Micro-syenite	Alkali feldspar (+ feldspathoid)	Medium-grained
Minnate	Biotite + orthoclase	Coarse-grained
Missourite	Pseudoleucite + aegirine + olivine	Coarse-grained
Monchiquite	Barkevikite/augite + feldspathoids	Coarse-grained
Monmouthite	Nepheline + albite + aegirine/Na-amphibole	Coarse-grained
Monzonite	Potash feldspar + plagioclase	Coarse-grained
Mugearite	Andesine/oligoclase bearing basalt	Fine-grained
Nepheline basanite	Alkali basalt with nepheline + olivine + alkali feldspar	Fine-grained
Nepheline-syenite	Nepheline + alkali feldspar	Coarse-grained
Nepheline-tephrite	Alkali basalt with nepheline + alkali feldspar	Fine-grained
Nephelinite	Alkali basalt with nepheline	Fine-grained
Nordmarkite	Microperthite + quartz	Fine-grained
Norite	Labradorite + orthopyroxenes (enstatite, hypersthene, bronzite)	Coarse-grained
Obsidian	Quartz + feldspar (sanidine, orthoclase)	Fine-grained
Oceanite	Olivine rich basalt	Fine-grained
Olivine leucitite	Alkali basalt with leucite + olivine	Fine-grained
Olivine nephelinite	Alkali basalt with nepheline + olivine	Fine-grained
Olivine-monzonite	Alkali feldspar + plagioclase + olivine	Fine-grained
Pegmatite	Any mineral	Very coarse-grained
Peridotite	Olivine + diopside	Coarse-grained
Perthosite	Perthite	Coarse-grained
Phonolite	Sanidine + hornblende + nepheline + biotite	Fine-grained
Picrite	Olivine + augite + calcic-plagioclase	Coarse-grained
Pulaskite	Antiperthite + aegirine + nepheline	Fine-grained
Pyroxenite	Pyroxene	Coarse-grained
Quartz-syenite	Quartz + alkali feldspar	Coarse-grained
Quartz-trachyte	Quartz + alkali feldspar	Fine-grained
Rhyodacite	Quartz + alkali feldspar = plagioclase	Fine-grained
Rhyolite	Quartz + alkali feldspar > plagioclase	Fine-grained
Serpentinite	Serpentine	Coarse-grained
Spessartite	Hornblende + plagioclase	Coarse-grained
Spilite	Albite-basalt	Fine-grained
Syenite	Alkali feldspar (+ feldspathoid)	Coarse-grained
Tachylite	Basaltic glass	Fine-grained
Teschenite	Labradorite + analsite + titan-augite + barkevikite	Coarse-grained
Theralite	Labradorite + nepheline + titan-augite + barkevikite	Coarse-grained
Tinguaite	Nepheline-syenite with sanidine + aegirine	Coarse-grained
Tonalite	Plagioclase+ quartz+ hornblende+ biotite	Coarse-grained
Trachy-andesite	Alkali feldspar + plagioclase	Fine-grained
Trachyte	Feldspar + quartz	Fine-grained
Troctolite	Labradorite + olivine	Coarse-grained
Trondjemite	Granodiorite with very less alkali feldspar	Coarse-grained
Ultrabasic-charnockite	Hypersthene + diopside + augite + hornblende	Coarse-grained
Vogesite	Hornblende + orthoclase	Coarse-grained
Wehrlite	Olivine + clino-pyroxene	Coarse-grained

TABLE 3.2: Description of common igneous rocks

Rock	Physical character	Texture	Essential minerals	Accessory minerals
Adamellite	Leucocratic, massive, hard and compact	Holocrystalline, phanero-crystalline, coarse-grained, hypidiomorphic, porphyritic	Potash feldspar (orthoclase and microcline) and plagioclase (oligoclase) in equal amounts, quartz	Sphene, pyroxene, hornblende, biotite, zircon, apatite, iron oxide, etc.
Andesite	Melanocratic, massive, hard and compact	Hemicrystalline, fine-grained, aphanitic, allotriomorphic, microporphyritic	Potash feldspar (orthoclase, perthite), plagioclase (oligoclase, andesine) with cryptocrystalline or glassy groundmass	Quartz, pyroxene (hypersthene, enstatite), olivine, biotite, hornblende, iron oxide, etc.
Anorthosite	Leucocratic, light to bluish gray, medium- to coarse-grained, massive, hard and compact	Holocrystalline, phanero-crystalline, coarse grained, equigranular, hypidiomorphic	Plagioclase (mostly labradorite, bytownite, anorthite), orthoclase and perthite in smaller amounts	Pyroxene, olivine or quartz, magnetite, ilmenite, rutile, etc.
Aplite	Leucocratic, massive, hard	Holocrystalline, fine-grained, equigranular, allotriomorphic	Potash feldspar, plagioclase and quartz	Pyroxene, hornblende, tourmaline, zircon, apatite, topaz, beryl, iron oxide, etc.
Basalt	Melanocratic, black, massive, hard, compact, vesicular, amygdaloidal	Holo- to hemi-crystalline, aphanitic, fine grained, poikilitic, ophitic to subophitic, intergranular, trachytic, vitrophyric	Plagioclase (mostly labradorite) and pyroxene	Olivine (rarely quartz), hypersthene, enstatite, hornblende, biotite, zeolite, magnetite, ilmenite
Carbonatite	Leucocratic, white, hard, compact, saccharoidal	Holocrystalline, phanero-crystalline, coarse-grained, equigranular, hypidiomorphic	Calcite, dolomite / ankerite	Apatite, biotite, pyroxene, magnesite, graphite
Charnockite	Melanocratic, bluish black, massive, hard, compact	Holocrystalline, phanero-crystalline, coarse-grained, hypidiomorphic	Hypersthene, quartz, feldspar (alkali and plagioclase)	Biotite, hornblende, rutile, zircon, ilmenite
Dacite	Leucocratic, massive, hard, compact	Flow texture, holo- to hemi-crystalline, porphyritic	Quartz phenocrysts and plagioclase groundmass	Hornblende and pyroxene
Diabase	Melanocratic, massive, hard, compact	Fine-grained, ophitic	Plagioclase (labradorite) and pyroxene (augite), often altered to amphibole	Zeolite, olivine, serpentine, hornblende, chlorite, ilmenite, leucoxene, etc.
Diorite	Melanocratic, massive, hard, compact	Holocrystalline, phanero-crystalline, medium- to coarse-grained, hypidiomorphic, rarely porphyritic/poikilitic.	Plagioclase (oligoclase to andesine), hornblende, biotite, potash feldspars may or may not be present	Sphene, feldspathoids, hypersthene, minor amount of quartz

(Contd...)

TABLE 3.2: Description of common igneous rocks (*Contd...*)

Rock	Physical character	Texture	Essential minerals	Accessory minerals
Dolerite	Mesocratic, greenish black, massive, hard, compact	Holocrystalline, phanerocrystalline, hypidiomorphic, medium-grained, porphyritic, poikilitic, ophitic, banding	Pyroxene (augite) and plagioclase (commonly labradorite)	Magnetite, hornblende, hypersthene, biotite, quartz (or olivine), chlorite, sphene, epidote, serpentine
Dunite	Melanocratic, yellowish green, massive, hard, compact	Holocrystalline, phanerocrystalline, hypidiomorphic, coarse-grained, poikilitic	Olivine	Serpentine, augite, plagioclase, spinel, chromite, talc, magnetite, ilmenite, etc.
Essexite	Mesocratic, massive, hard, compact	Holocrystalline, phanerocrystalline, coarse-grained, hypidiomorphic	Plagioclase (labradorite), orthoclase, augite, hornblende, biotite	Feldspathoids (nepheline, sodalite, cancrinite)
Eclogite	Melanocratic, massive, hard, compact, may be banded, heavy	Holocrystalline, phanerocrystalline, hypidiomorphic, coarse-grained, porphyritic, poikilitic, reaction structure	Garnet (pyrope, almandine), omphacite	Chrome-diopside, kyanite, quartz or olivine, sphene, rutile, corundum
Felsite	Leucocratic, massive, hard, compact	Hemicrystalline, aphanitic, felsophyric to vitrophyric	Potash and plagioclase feldspars	Quartz, etc.
Gabbro	Melanocratic, hard, compact	Holocrystalline, phanerocrystalline, coarse-grained, hypidiomorphic, porphyritic, poikilitic, orbicular structure, banding	Labradorite, clinopyroxenes (augite, diopside, pigeonite), olivine	Hornblende, biotite, ilmenite, magnetite, olivine, less commonly quartz, sphene, epidote, serpentine
Granite	Leucocratic, pinkish or whitish in colour, massive, compact, hard and may be jointed	Holocrystalline, phanerocrystalline, medium- to coarse-grained, hypidiomorphic, porphyritic, quartz and feldspar may show intergrowth (graphic) texture	Quartz, potash feldspar (orthoclase and microcline) dominates the plagioclase (more sodic, oligoclase) and perthite	Pyroxene (augite and hypersthene), hornblende, garnet, tourmaline, biotite, muscovite, zircon, apatite, topaz, iron oxide, etc.
Granodiorite	Leucocratic, massive, hard, compact	Holocrystalline, phanerocrystalline, medium- to coarse-grained, hypidiomorphic	Quartz, plagioclase (oligoclase, andesine), dominate over potash feldspar (orthoclase)	Hornblende, pyroxene (augite and hypersthene), biotite, zircon, apatite, ilmenite, magnetite, sphene
Granophyre	Leucocratic, massive, hard, compact	Holocrystalline, phanerocrystalline, medium- to coarse-grained, hypidiomorphic, micrographic	Quartz, potash feldspar	Plagioclase (albite, oligocene), biotite, apatite, amphibole, zircon, etc.

(*Contd...*)

TABLE 3.2: Description of common igneous rocks (*Contd...*)

Rock	Physical character	Texture	Essential minerals	Accessory minerals
Hornblendite	Melanocratic, hard, compact	Holocrystalline, phanero-crystalline, coarse-grained, hypidiomorphic	Hornblende	Pyroxene, olivine, chromite, ilmenite, magnetite
Kimberlite	Melanocratic, brecciated, hard	Holocrystalline, phanero-crystalline, medium- to coarse-grained, panidiomorphic to hypidiomorphic	Phlogopite, bronzite, chrome-diopside	Diamond, apatite, garnet, chromite, ilmenite, olivine, plagioclase, etc.
Lamprophyre	Melanocratic, massive, hard	Holocrystalline, phanero-crystalline, medium- to coarse-grained, panidiomorphic, porphyritic, reaction rim and corona	Biotite/phlogopite, horn-blende, augite/barkevikite, orthoclase, plagioclase (oligoclase, andesine),	Chlorite, apatite, sphene, zircon, rarely quartz
Monzonite	Leucocratic, massive, hard	Holocrystalline, medium-grained, hypidiomorphic, porphyritic	Orthoclase, microcline, sodic plagioclase	Quartz, hornblende, biotite, sphene, rutile
Nepheline-syenite	Leucocratic, massive, hard, compact, greasy lustre	Holocrystalline, phanero-crystalline, medium- to coarse-grained, hypidiomorphic, porphyritic	Nepheline, orthoclase, micro-cline, albite, micro-perthite	Hornblende, pyroxene, biotite
Norite	Melanocratic, hard, compact	Holocrystalline, phanero-crystalline, coarse-grained, hypidiomorphic, porphyritic, poikolitic, orbicular structure, banding	Labradorite, orthopyroxenes (enstatite, hypersthene, bronzite)	Hornblende, olivine, biotite, hematite, magnetite, sphene, epidote, serpentine
Obsidian	Melanocratic, massive, compact, hard, low speci-fic gravity, conchoidal fracture	Holohyaline, aphanitic, fine-grained	Quartz, cristobalite, tridimite, feldspar (sanidine, orthoclase)	Oligoclase, pyroxene, horn-blende, magnetite
Pegmatite (Granite pegmatite)	Leucocratic, massive, compact, hard	Holocrystalline, pegmatitic, phanerocrystalline, very coarse-grained, panidiomorphic to hypidiomorphic, occasionally graphic	Quartz, potash feldspar (orthoclase, microcline), biotite, muscovite, tourmaline are common; garnet and other minerals may be present	Albite, beryl, topaz, etc.

(*Contd...*)

TABLE 3.2: Description of common igneous rocks (*Contd...*)

Rock	Physical character	Texture	Essential minerals	Accessory minerals
Peridotite	Melanocratic, massive, compact, hard, saccharoidal, banded	Holocrystalline, phanerocrystalline, coarse-grained, hypidiomorphic, ophitic, reaction rim	Olivine, diopside	Hornblende, serpentine, ilmenite, magnetite
Phonolite	Melanocratic, massive, compact, hard	Hemicrystalline to holohyaline, trachytic	Sanidine, hornblende, nepheline, biotite	Leucite, sodalite, cancrinite
Picrite	Melanocratic, massive, compact, hard	Hemicrystalline, aphanitic, porphyritic, poikilitic, ophitic to subophitic	Olivine, pyroxene (augite, diopside, hypersthene, enstatite, bronzite) plagioclase (labradorite, bytwonite), biotite	Chromite, magnetite, ilmenite, sphene, rutile, serpentine, etc.
Pitchstone	Melanocratic, massive, compact, hard	Holohyaline, aphanitic, fine-grained	Quartz and feldspar	Sanidine, oligoclase, pyroxene, hornblende, magnetite, etc.
Pumice	Mesocratic, soft, light	Holohyaline, aphalitic		Zeolite, calcite, etc.
Pyroxenite	Melanocratic, massive, compact, hard	Holocrystalline, phanerocrystalline, coarse-grained, hypidiomorphic, poikilitic	Any one or two of the pyroxenes—augite, enstatite, hypersthene, bronzite, pigeonite	Olivine, hornblende, biotite, phlogopite, plagioclase (labradorite, bytwonite), chromite, magnetite, rarely quartz
Rhyolite	Leucocratic, hard, compact, flow structure	Hemicrystaline, aphanitic, fine-grained, microporphyritic	Quartz, crystobalite, tridimyte, sanidine	Feldspar, pyroxene, hornblende, biotite, magnetite, etc.
Serpentinite	Melanocratic, massive, soft	Holocrystalline, phanerocrystalline, hypidiomorphic, coarse-grained, poikilitic	Serpentine (antigorite, chrysotile, lizardite)	Pyroxene, olivine
Syenite	Leucocratic, massive, hard, compact	Holocrystalline, phanerocrystalline, medium-to coarse-grained, hypidiomorphic, porphyritic	Orthoclase, microcline, albite, perthite	Hornblende, biotite, plagioclase, nepheline, pyroxene, opaque minerals, little or no quartz, sphene, rutile
Tonalite	Leucocratic, massive, hard	Holocrystalline, phanerocrystalline, medium- to coarse-grained, hypidiomorphic	Plagioclase (oligoclase, andesine), quartz, hornblende, biotite	Orthoclase, apatite, sphene, zircon, ilmenite, magnetite
Trachyte	Mesocratic, massive, hard, compact	Holocrystalline, phanerocrystalline, fine- to medium-grained, hypidiomorphic, porphyritic, trachitic	Feldspar (orthoclase, plagioclase) and quartz embedded in a fine-grained groundmass	Hornblende, pyroxene, nepheline (occasional)

4

Calculation of Norm

Calculation of norm is the method of expressing the chemical composition of an igneous rock in terms of a series of arbitrarily selected minerals, according to a set of prescribed rules. Four American petrologists, Cross, Iddings, Pirsson and Washington proposed a classification scheme of igneous rocks based on their chemical composition. CH Kelsey devised a set of rules to calculate the normative mineral composition of igneous rocks from the weight percent of oxides, which constitute the rock. These rules are based on the findings of experimental petrology made in the laboratory from silicate melt of known chemical composition. The normative minerals are anhydrous. Thus, the norm of a rock containing appreciable amounts of hydrated and hydroxyl-bearing minerals will show significant deviation from its mode (actual mineral content). The formula weights of oxides and elements, which constitute the common igneous rocks are given in Table 4.1.

4.1 RULES FOR CALCULATION OF NORM

The oxide constituents are allocated in a definite order to form normative minerals. The allocation is similar to sequential crystallization of minerals from magmas. However, normative minerals do not contain water, complete solid solution is not permitted (pyroxenes do not contain Al_2O_3, Fe_2O_3 or TiO_2), and normative minerals are independent of composition of magma. The normative minerals are given in Table 4.2.

TABLE 4.1: Molecular weights of oxides and atomic weight of elements
(*Rounded to nearest integer for simplicity of calculation*)

Oxide	Wt.	Oxide	Wt.	Oxide	Wt.	Oxide	Wt.	Element	Wt.
SiO_2	60	MgO	40	CO_2	44	MnO	71	S	32
Al_2O_3	102	CaO	56	TiO_2	80	NiO	75	Cl	35.5
Fe_2O_3	160	Na_2O	62	P_2O_5	142	BaO	153	F	19
FeO	72	K_2O	94	SO_3	80				
Cr_2O_3	152	H_2O	18	ZrO_2	123				

TABLE 4.2: Composition and formula weights of normative minerals
(*Rounded to nearest integer for simplicity of calculation*)

Normative mineral	Composition	Formula weight
Acmite (Ac)	$Na_2O.Fe_2O_3.4SiO_2$	462
Albite (Ab)	$Na_2O.Al_2O_3.6SiO_2$	524
Anorthite (An)	$CaO.Al_2O_3.2SiO_2$	278
Apatite (Ap)	$3\,CaO.P_2O_5.\frac{1}{3}CaF_2$	336
Calcite (Cc)	$CaO.CO_2$	100
Cancrinite (Nc)	$Na_2O.CO_2$	106
Chromite (Cm)	$FeO.Cr_2O_3$	224
Corundum (C)	Al_2O_3	102
Dicalcium silicate (Cs)*	$2CaO.SiO_2$	172
Diopside (Di)	$CaO.(Mg, Fe)O.2SiO_2$	116, 100, 132
Fluorite (Fr)	CaF_2	78
Halite (Hl)	$NaCl$	58
Hematite (Hm)	Fe_2O_3	160
Hypersthene (Hy)	$(Mg, Fe)O.SiO_2$	100, 132
Ilmenite (Il)	$FeO.TiO_2$	152
Kaliophillite (Kp)	$K_2O.Al_2O_3.2SiO_2$	316
Leucite (Lc)	$K_2O.Al_2O_3.4SiO_2$	436
Magnetite (Mt)	$FeO.Fe_2O_3$	232
Nepheline (Ne)	$Na_2O.Al_2O_3.2SiO_2$	284
Olivine (Ol)	$2(Mg, Fe)O.SiO_2$	140, 204
Orthoclase (Or)	$K_2O.Al_2O_3.6SiO_2$	556
Perovskite (Pf)	$CaO.TiO_2$	136
Potassium metasilicate (Ks)	$K_2O.SiO_2$	154
Pyrite (Pr)	FeS_2	120
Quartz (Q)	SiO_2	60
Rutile (Ru)	TiO_2	80
Sodium metasilicate (Ns)	$Na_2O.SiO_2$	122
Sphene (titanite) (Tn)	$CaO.TiO_2.SiO_2$	196
Thenardite (Th)	$Na_2O.SO_3$	142
Wollastonite (Wo)	$CaO.SiO_2$	116
Zircon (Z)	$ZrO_2.SiO_2$	183

*Dicalcium silicate (Cs) is known as 'larnite', a mineral crystallising in monoclinic system

The CIPW norm form is generally used for calculation of normative composition of igneous rocks. The molecular weights of common oxides and minerals are provided in the norm form (Fig. 4.1). The step-by-step rules for calculation of norm are given below.

1. Calculate the molecular proportions of different oxides by dividing the weight percentages by respective formula weights. For example, if the weight percentages of SiO_2, Al_2O_3 and Fe_2O_3 are 69.92, 13.91 and 1.18 respectively, then their molecular proportions are (69.92 ÷ 60), (13.91 ÷ 102) and (1.18 ÷ 160), i.e. 1.165, 0.136 and 0.007 respectively.

2. Add weight percentages of (MnO + NiO) to FeO and (BaO + SrO) to CaO.

Constituents of rock	SiO$_2$	Al$_2$O$_3$	Fe$_2$O$_3$	FeO	MgO	CaO	Na$_2$O	K$_2$O	H$_2$O	CO$_2$	TiO$_2$	P$_2$O$_5$	SO$_3$	S	Cl	F	MnO	Molecular proportions	Molecular weights	Percentages (norm)	Group of standard minerals
Percentages (analysis)																					
Molecular weights	60	102	160	72	40	56	62	94	18	44	80	142	80	64	71	38	71				
Molecular proportions																					
Quartz SiO$_2$																			60		Q=
Orthoclase K$_2$O.Al$_2$O$_3$.6SiO$_2$																			556		F=
Albite Na$_2$O.Al$_2$O$_3$.6SiO$_2$																			524		
Anorthite CaO Al$_2$O$_3$.2SiO$_2$																			278		L=
Leucite K$_2$O.Al$_2$O$_3$.4SiO$_2$																			436		
Nepheline Na$_2$O.Al$_2$O$_3$.2SiO$_2$																			284		
Corundum Al$_2$O$_3$																			102		C=
Acmite Na$_2$O.Fe$_2$O$_3$.4SiO$_2$																			462		
Diopside CaO.SiO$_2$																			116		P=
Diopside MgO.SiO$_2$																			100		
Diopside FeO.SiO$_2$																			132		
Wollastonite CaO.SiO$_2$																			116		
Hypersthene MgO.SiO$_2$																			100		
Hypersthene FeO.SiO$_2$																			132		
Olivine 2MgO.SiO$_2$																			140		O=
Olivine 2FeO.SiO$_2$																			204		
Magnetite FeO.Fe$_2$O$_3$																			232		M=
Hematite Fe$_2$O$_3$																			160		
Ilmenite FeO.TiO$_2$																			152		
Pyrite FeS$_2$																			120		A=
Apatite 3CaO.P$_2$O$_5$⅓CaF$_2$																			336		
Calcite CaO.CO$_2$																			100		

Salic group = Q=, F=, L=, C=

Femic group = P=, O=, M=, A=

Fig. 4.1: CIPW norm form

3. Set zircon (Z) = ZrO_2 and allocate equal amount of SiO_2 to it.

4. Set apatite (Ap) = P_2O_5 and allocate $3.33 \times P_2O_5$ of CaO to Ap (or if F is present, allocate $3 \times P_2O_5$ of CaO and $0.33 \times F$). If, $F > 0.33 \times Ap$, allocate $0.33 \times F$ to Ap. If $F < 0.33 \times Ap$, all the F is to be allocated to Ap. If F is left over, set Fluorite (Fr) = remaining F and allocate $2 \times CaO$ to Fr.

5. Set halite (Hl) = Cl and allocate $(0.5 \times Cl)$ of Na_2O.

6. If SO_3 is calculated as SO_3 and not as S, set thenardite (Th) = SO_3; allocate equal amount of Na_2O to Th. (This is applicable if the rock contains hauyn).

7. If S is present or SO_3 is calculated as S, set pyrite (Pr) = S and allocate $(0.5 \times S)$ of FeO to Pr.

8. If the rock contains cancrinite (Nc), set Nc = CO_2 and allocate equal amount of Na_2O to Nc.

9. If the rock contains calcite (Cc), set Cc = CO_2 and allocate equal amount of CaO to Cc. (If the modal calcite is secondary or from associated limestone, it is not included in the norm)

10. Set chromite (Cm) = Cr_2O_3 and allocate equal amount of FeO to Cm.

11. If FeO > TiO_2, set Ilmenite (Il) = TiO_2 and allocate equal amount of FeO to Il. If FeO < TiO_2, set Il = FeO and allocate equal amount of TiO_2 to Il.

12. If $Al_2O_3 > K_2O$, set orthoclase (Or) = K_2O and allocate equal amount of Al_2O_3 and 6 times that of SiO_2. If $Al_2O_3 < K_2O$, set Or = Al_2O_3, and allocate equal amount of K_2O and 6 times that of SiO_2. The remaining K_2O and equal amount of SiO_2 should be allocated to potassium metasilicate (Ks).

13. If the remaining $Al_2O_3 > Na_2O$, set albite (Ab) = Na_2O and allocate equal amount of Al_2O_3 and 6 times that of SiO_2. If $Al_2O_3 < Na_2O$, set albite (Ab) = Al_2O_3 and allocate equal amount of Na_2O and 6 times that of SiO_2.

14. If $Fe_2O_3 >$ remaining Na_2O, set acmite (Ac) = Na_2O and allocate equal amount of Fe_2O_3 and 4 times that of SiO_2. If $Fe_2O_3 < Na_2O$, set acmite (Ac) = Fe_2O_3 and allocate equal amount of Na_2O and 4 times that of SiO_2. The remaining Na_2O is to be set as sodium metasilicate (Ns) and equal amount of SiO_2 should be allocated to it.

15. If remaining $Al_2O_3 > CaO$, set anorthite (An) = CaO and allocate equal amount of Al_2O_3 and 2 times that of SiO_2. The remaining Al_2O_3 should be allocated to corundum (C). If the available $Al_2O_3 < CaO$, set anorthite (An) = Al_2O_3 and allocate equal amount of CaO and 2 times that of SiO_2 to it.

16. If the remaining CaO > remaining TiO_2, set sphene (titanite) (Tn) = TiO_2 and allocate equal amount of CaO and SiO_2 to it. If the remaining CaO < remaining TiO_2, set sphene (Tn) = CaO and allocate equal amount of TiO_2 and SiO_2. Set the balance TiO_2 as rutile (Ru).

17. If $Fe_2O_3 >$ FeO, set magnetite (Mt) = FeO and allocate equal amount of Fe_2O_3. The remaining Fe_2O_3 is to be set as hematite (Hm). If $Fe_2O_3 <$ FeO, set magnetite (Mt) = Fe_2O_3 and allocate equal amount of FeO to it.

18. Determine $[MgO \div (MgO + FeO)] = a$ and $[FeO \div (MgO + FeO)] = b$ from balance amount of MgO and FeO. These parameters are to be used for allocation of MgO and FeO to diopside, hypersthene and olivine.

19. If the remaining CaO > (MgO + FeO) set diopside (Di) = (MgO + FeO) and allocate equal amount of CaO and (CaO + MgO + FeO) amount of SiO_2. Set wollastonite (Wo) = remaining CaO and allocate equal amount of SiO_2 to it. If the remaining CaO < (MgO + FeO) set diopside (Di) = CaO and allocate equal amount of (MgO + FeO) in the ratio a and b

respectively and $(CaO + MgO + FeO)$ amount of SiO_2. The remaining MgO and FeO should be set as hypersthene (Hy) in the ratio a:b and an amount of SiO_2 equal to $(MgO + FeO)$ should be allocated to it.

20. Now the entire amount of SiO_2 allocated to different minerals is to be added (say it is Y). If $SiO_2 > Y$ set quartz $(Q) = SiO_2 - Y$. The calculation of norm is complete at this stage. Multiply the amount of each normative mineral by respective formula (molecular) weight to find out the percentage of each normative mineral. If $SiO_2 < Y$ then proceed further to compensate the deficiency $(D = Y - SiO_2)$.

21. If $D < Hy/2$, set olivine $(Ol) = D$ [MgO and FeO in ratio a:b and SiO_2 equal to $(MgO + FeO)/2$] and $Hy = Hy - 2D$. It is to be noted here that normative molecular proportion of Ol is equal to $(MgO + FeO)/2$, i.e. equal to the amount of allocated silica. The silica deficiency is now zero. If $D > Hy/2$, set $Ol = Hy$ and $Hy = 0$; $D_1 = D - Hy/2$.

22. If $D_1 < Tn$, set $Tn = Tn - D_1$ and $Pf = D_1$. The silica deficiency is now zero. If $D_1 > Tn$, set $Tn = Pf$ and $Tn = 0$; $D_2 = D_1 - Tn$.

23. If $D_2 < 4Ab$, set nepheline $(Ne) = D_2/4$ [$Na_2O = Al_2O_3 = D_2/4$ and $SiO_2 = D_2/2$] and $Ab = Ab - D_2/4$. The silica deficiency is now zero. If $D_2 > 4Ab$, set $Ne = Ab$ and $Ab = 0$, $D_3 = D_2 - 4Ab$.

24. If $D_3 < 2Or$, set leucite $(Lc) = D_3/2$ [$K_2O = Al_2O_3 = D_3/2$ and $SiO_2 = 2D_3$] and $Or = Or - D_3/2$. The silica deficiency is now zero. If $D_3 > 2Or$, set $Lc = Or$ and $Or = 0$; $D_4 = D_3 - 2Or$.

25. If $D_4 < Wo/2$, set dicalcium silicate $(Cs) = D_4$ [$CaO = 2D_4$ and $SiO_2 = D_4$] and $Wo = Wo - 2D_4$. The silica deficiency is now zero. If $D_4 < Wo/2$, set $Cs = Wo$ and $Wo = 0$; $D_5 = D_4 - Wo/2$.

26. If $D_5 < Di$ add an amount of $D_5/2$ to the amounts of Cs and Ol already in the norm; set $Di = Di - D_5$. The silica deficiency is now zero. If $D_5 > Di$, add an amount of $Di/2$ to the amounts of Cs and Ol already in the norm, put $Di = 0$; $D_6 = D_5 - Di$.

27. Set kaliophillite $(Kp) = D_6/2$ [$K_2O = Al_2O_3 = D_6/2$ and $SiO_2 = D_6$] and $Lc = Lc - D_6/2$ [$K_2O = Al_2O_3 = D_6/2$ and $SiO_2 = 2D_6$]. The silica deficiency is now zero.

[While compensating the deficiency, adjustment is made in the molecular proportion of allocated SiO_2, keeping the molecular proportion of other oxides fixed.]

Now the calculation of norm is complete. Multiply the amount of each normative mineral by respective formula weight given in Table 4.2 to find out the percentage of each normative mineral. The computed values of normative minerals are used for classification of the rocks into class, order and rangs. The classification is outlined below.

Salic group

      Q = quartz; F = orthoclase + albite + anorthite; L = leucite + nepheline + kaliophilite;

      C = corundum, Z = zircon + halite + thenardite

Femic group

      P = acmite + diopside + wollastonite + hypersthene + sodium metasilicate + potassium metasilicate

      O = olivine + dicalcium silicate

      H = hematite + magnetite + chromite

      T = ilmenite + sphene + rutile + perovskite

      M = H + T

      A = apatite + fluorite + pyrite + calcite + cancrinite

4.2 NORMATIVE CLASSIFICATION OF IGNEOUS ROCKS

The igneous rocks are classified into different classes, orders, rangs and subrangs.

  i. Division into different classes is made on the basis of salic/femic ratio.

Class	Salic/femic ratio
I. Persalic	> 7.0
II. Dosalic	7.0 – 1.667
III. Salfemic	1.667 – 0.60
IV. Dofemic	0.60 – 0.143
V. Perfemic	< 0.143

  ii. Each of the classes I, II and III are divisible into 9 orders on the basis of relative amounts of Q, F and L. The classes IV and V are divided into 5 orders each depending on the amount of M and P + O.

Orders of classes I, II and III		Orders of classes IV and V	
1. Perquaric	$Q \gg F$	1. Perpolic	$P + O \gg M$
2. Feldoquaric	$Q > F$	2. Dopolic	$P + O > M$
3. Quarfelic	$Q = F$	3. Polmitic	$P + O = M$
4. Quardofelic	$Q < F$	4. Domitic	$P + O < M$
5. Perfelic	$Q \text{ or } L \ll F$	5. Permitic	$P + O \ll M$
6. Lendofelic	$F > L$		
7. Lenfelic	$F = L$		
8. Feldolenic	$F < L$		
9. Perlenic	$F \ll L$		

  iii. The classes I, II and III are divided into 5 rangs each on the basis of relative proportion of salicalkali ($Na_2O + K_2O$) and saliclime (CaO). Each of the classes IV and V are divisible into 5 rangs on the basis of relative proportion of total magnesia, ferrous oxide and femicalkali to femiclime.

Rangs of classes I, II and III		Rangs of classes IV and V	
1. Peralkalic	$(Na_2O + K_2O) \gg CaO$	1. Permiric	$(MgO + FeO + Na_2O) \gg CaO$
2. Doalkalic	$(Na_2O + K_2O) > CaO$	2. Domiric	$(MgO + FeO + Na_2O) > CaO$
3. Alkalicalcic	$(Na_2O + K_2O) = CaO$	3. Calcimiric	$(MgO + FeO + Na_2O) = CaO$
4. Docalcic	$(Na_2O + K_2O) < CaO$	4. Docalcic	$(MgO + FeO + Na_2O) < CaO$
5. Percalcic	$(Na_2O + K_2O) \ll CaO$	5. Percalcic	$(MgO + FeO + Na_2O) \ll CaO$

  iv. Rangs 1, 2 and 3 of classes I, II and III are divisible into 5 subrangs on the basis of relative proportion of Na_2O and K_2O. These are:

Subrangs of Rangs 1, 2 and 3		Subrangs of Rang 4	
1. Perpotassic	$K_2O \gg Na_2O$	1–2. Perpotassic	$K_2O > Na_2O$
2. Dopotassic	$K_2O > Na_2O$	3. Sodipotassic	$K_2O = Na_2O$
3. Sodipotassic	$K_2O = Na_2O$	4–5. Persodic	$K_2O < Na_2O$
4. Dosodic	$K_2O < Na_2O$		
5. Persodic	$K_2O \ll Na_2O$	Rang 5 has no subrang	

v. Rangs 1, 2 and 3 of classes IV and V are divisible into 5 subrangs on the basis of relative proportion of MgO and FeO + Na$_2$O. These are:

Subrangs of Rangs 1, 2 and 3		Subrangs of Rang 4	
1. Permagnesic	MgO >> FeO + Na$_2$O	1–2. Permagnessic	MgO > FeO
2. Domagnesic	MgO > FeO + Na$_2$O	3. Magnesiferrous	MgO = FeO
3. Magnesiferrous	MgO = FeO + Na$_2$O	4–5. Perferrous	MgO < FeO
4. Doferrous	MgO < FeO + Na$_2$O		
5. Perferrous	MgO << FeO + Na$_2$O	Rang 5 has no subrang	

4.3 CALCULATION OF NIGGLI VALUES AND OTHER PARAMETERS

The weight percentages of different major element oxides are calculated as molecular numbers (molecular proportion) by dividing the percent values by respective molecular weights as done in the first step of norm calculation.

The molecular number of Fe$_2$O$_3$ is multiplied by 2 and is added to the value of FeO. Thus, the total iron is expressed in terms of molecular number of FeO, i.e. FeO$_t$ (FeO$_t$ = (2 × Fe$_2$O$_3$ + FeO)). The sum of molecular numbers of (Al$_2$O$_3$ + Cr$_2$O$_3$), (FeO$_t$ + MgO + MnO), CaO and (Na$_2$O + K$_2$O) are recalculated to 100 and are expressed by 'al', 'fm', 'c' and 'alk' respectively.

i. $\text{al} = \dfrac{(Al_2O_3 + Cr_2O_3)}{(Al_2O_3 + Cr_2O_3 + 2 \times Fe_2O_3 + FeO + MgO + MnO + CaO + Na_2O + K_2O)} \times 100$

ii. $\text{fm} = \dfrac{(2 \times Fe_2O_3 + FeO + MgO + MnO)}{(Al_2O_3 + Cr_2O_3 + 2 \times Fe_2O_3 + FeO + MgO + MnO + CaO + Na_2O + K_2O)} \times 100$

iii. $\text{c} = \dfrac{CaO}{(Al_2O_3 + Cr_2O_3 + 2 \times Fe_2O_3 + FeO + MgO + MnO + CaO + Na_2O + K_2O)} \times 100$

iv. $\text{alk} = \dfrac{(Na_2O + K_2O)}{(Al_2O_3 + Cr_2O_3 + 2 \times Fe_2O_3 + FeO + MgO + MnO + CaO + Na_2O + K_2O)} \times 100$

The molecular number of SiO$_2$ is reduced in the same proportion and indicated by 'si'.

v. $\text{si} = \dfrac{SiO_2}{(Al_2O_3 + Cr_2O_3 + 2 \times Fe_2O_3 + FeO + MgO + MnO + CaO + Na_2O + K_2)} \times 100$

The recalculated values of 'al', 'fm', 'c', 'alk' and 'si' are known as Niggli values.

vi. The 'qz' value is calculated by the formula: qz = si – (100 + 4 alk).

The qz value is positive for oversaturated rocks and negative for under saturated rocks. Values of 'k' and 'mg' are determined by the following formulae where all oxides are in molecular proportion

vii. $\text{k} = \dfrac{K_2O}{(Na_2O + K_2O)} \times 100$

viii. $\text{mg} = \dfrac{MgO}{(2 \times Fe_2O_3 + FeO + MgO + MnO)} \times 100$

ix. Differentiation Index (DI) is the sum of normative quartz (Q), orthoclase (Or), albite (Ab), nepheline (Ne), leucite (Lc) and potassium metasilicate (Ks).

$$DI = Q + Or + Ab + Ne + Lc + Ks$$

x. Fractionation Index (FI) is the sum of normative quartz (Q), orthoclase (Or), albite (Ab), acmite (Ac) and sodium metasilicate (Ns).

$$FI = Q + Or + Ab + Ac + Ns$$

xi. Crystallisation index (CI) = Anorthite + (2.157003 × Mg-diopside) + (0.700837 × Mg-hypersthene) + Mg-olivine.

(Crystallisation index is calculated after recalculating the norm to 100%)

xi. Colour index = Total normative mafics = Di + Hy + Ol + Mt + Il + Ac.

xiii. Agpaitic index = $\dfrac{(Na_2O + K_2O)}{Al_2O_3}$

xiv. Magnesium number (M′) = $\dfrac{MgO}{(MgO + FeO)} \times 100$

(For Agpaitic index and magnesium number, all oxides are in molecular proportion)

xv. Solidification index (SI) = $\dfrac{MgO}{(MgO + FeO + Fe_2O_3 + Na_2O + K_2O)} \times 100$

xvi. Felsic index (FI) = $\dfrac{(Na_2O + K_2O)}{(CaO + Na_2O + K_2O)} \times 100$

xvii. Mafic index (MI) = $\dfrac{(FeO + Fe_2O_3)}{(MgO + FeO + Fe_2O_3)} \times 100$

xviii. Oxidation ratio (OR) = $\dfrac{2Fe_2O_3}{(2FeO + Fe_2O_3)} \times 100$

(For SI, FI, MI and OR, all oxides are in weight percent).

4.4 SIGNIFICANCE OF NIGGLI VALUES

Some petrologists use the Niggli values (c, fm, al and alk) for classification of igneous rocks. As it does not convey any idea regarding the texture and mineralogy of the rocks, it is a classification of magmas rather than the igneous rocks in true sense. The classification scheme is illustrated in the form of a tetrahedron shown in Fig. 4.2.

Four apices of the tetrahedron are occupied by al, alk, c and fm. The al-alk-fm face (right to the observer) is moved about the al-alk line as a hinge till it meets the al-alk-c face. In the process, the c-fm line is divided into ten equal segments (I to X). The segment I lies between c_0-fm_{100} to c_{10}-fm_{90}, segment II lies between c_{10}-fm_{90} to c_{20}-fm_{80} and so on. The rock is classified according to the segment in which the composition is plotted.

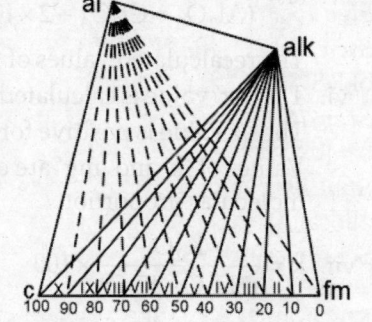

Fig. 4.2: Niggli's tetrahedron showing the segments of rock compositions

4.5 NORM CALCULATION OF AN OVERSATURATED ROCK

The chemical analysis of an oversaturated igneous rock is given in Table 4.3. The normative mineral composition is to be calculated and the Niggli values and other parameters are to be determined.

The step-by-step procedure of calculation of norm is given below and the completed norm form is shown in Fig. 4.3.

TABLE 4.3: Oxide composition of an oversaturated igneous rock												
Oxide	SiO_2	Al_2O_3	Fe_2O_3	FeO	MgO	CaO	Na_2O	K_2O	TiO_2	P_2O_5	CO_2	MnO
Wt.%	58.90	15.50	2.10	4.50	5.10	6.20	3.60	2.50	1.05	0.25	0.15	0.15

Step 1. Convert the weight percentages to molecular proportions (mol. prop.) by dividing the weight percentages by respective molecular weights.

Step 2. Add MnO (0.002) to FeO (0.063), So FeO = 0.065.

Step 3 is not applicable as ZrO_2 is absent.

Step 4. Set apatite = P_2O_5 (0.002) and allocate 3.33 × 0.002 = 0.007 CaO. F is not available. So **apatite = 0.002** and balance CaO = 0.111 – 0.007 = 0.104.

Steps 5–8 are not applicable.

Step 9. Set Cc = CO_2 (0.003) and allocate 0.003 of CaO to Cc. So **calcite = 0.003** and balance CaO = 0.104 – 0.003 = 0.101.

Step 10 is not applicable as Cr_2O_3 is absent.

Step 11. FeO = 0.065 and TiO_2 = 0.013. Set ilmenite = TiO_2 (0.013) and allocate FeO = 0.013 to it. So **ilmenite = 0.013.** TiO_2 = 0 and balance FeO = 0.065 – 0.013 = 0.052.

Step 12. Al_2O_3 = 0.152 and K_2O = 0.027. Set orthoclase = K_2O (0.027) and allocate Al_2O_3 = 0.027 and SiO_2 = 6 × 0.027 = 0.162 to it. K_2O is exhausted and balance Al_2O_3 = 0.152 – 0.027 = 0.125. So **orthoclase = 0.027**.

Step 13. Balance Al_2O_3 (0.125) > Na_2O (0.058). Set **albite = 0.058;** allocate equal amounts of Na_2O, Al_2O_3 and (6 × 0.058 = 0.348) of SiO_2. Na_2O is exhausted and balance Al_2O_3 = 0.125 – 0.058 = 0.067.

Step 14 is not applicable as Na_2O = 0.

Step 15. Balance Al_2O_3 (0.067) < balance CaO (0.101). Set **anorthite = 0.067** and allocate equal amounts of CaO, Al_2O_3 and (2 × 0.067 = 0.134) of SiO_2. Al_2O_3 is exhausted and CaO balance is 0.034.

Step 16 is not applicable

Step 17. FeO = 0.052 (step 12) > Fe_2O_3 = 0.013. Set **magnetite = 0.013** (= Fe_2O_3) and allocate 0.013 of FeO. Fe_2O_3 is exhausted and balance FeO is (0.052 – 0.013) 0.039.

Step 18. Balance FeO = 0.039 and MgO = 0.128. a = [0.128 ÷ (0.128 + 0.039)] = 0.77 and b = [0.039 ÷ (0.128 + 0.039)] = 0.23.

Step 19. Balance CaO = 0.034 (step 15) and MgO = 0.128. Set Diopside = 0.034 (= CaO) and allocate 0.026 MgO (= 0.77 × 0.034), 0.008 FeO (= 0.23 × 0.034) and (0.034 + 0.026 + 0.008) 0.068 SiO_2. Set remaining MgO (0.128 – 0.026 = 0.102) as Mg-hypersthene and balance FeO (0.039 – 0.008 = 0.031) as Fe-hypersthene and allocate (0.102 + 0.031) 0.133 SiO_2 to them. So **Ca-diopside = 0.034, Mg-diopside = 0.026, Fe-diopside = 0.008, Mg-hypersthene = 0.102.** and **Fe-hypersthene = 0.031.** CaO and MgO are exhausted.

Constituents of rock	SiO₂	Al₂O₃	Fe₂O₃	FeO	MgO	CaO	Na₂O	K₂O	H₂O	CO₂	TiO₂	P₂O₅	SO₃	S	Cl	F	MnO	Molecular proportions	Molecular weights	Percentages (norm)	Group of standard minerals
Percentages (analysis)	58.90	15.5	2.10	4.50	5.10	6.20	3.60	2.50		0.15	1.05	0.25					0.15				
Molecular weights	60	102	160	72	40	56	62	94	18	44	80	142	80	64	71	38	71				
Molecular proportions	0.982	0.152	0.013	0.063	0.128	0.111	0.058	0.027		0.003	0.013	0.002					0.002				
Quartz — SiO₂	0.137																	0.137	60	8.22	Q = 8.22
Orthoclase — K₂O.Al₂O₃.6SiO₂	0.162	0.027						0.027										0.027	556	15.01	F = 64.03
Albite — Na₂O.Al₂O₃.6SiO₂	0.348	0.058					0.058											0.058	524	30.39	
Anorthite — CaO Al₂O₃.2SiO₂	0.134	0.067				0.067												0.067	278	18.36	
Leucite — K₂O.Al₂O₃.4SiO₂																			436		L =
Nepheline — Na₂O.Al₂O₃.2SiO₂																			284		
Corundum — Al₂O₃		0.067																	102		C =
Acmite — Na₂O.Fe₂O₃.4SiO₂																			462		
Diopside — CaO.SiO₂	0.034					0.034												0.034	116	3.94	P = 21.89
Diopside — MgO.SiO₂	0.026				0.026													0.026	100	2.60	
Diopside — FeO.SiO₂	0.008			0.008														0.008	132	1.06	
Wollastonite — CaO.SiO₂																			116		
Hypersthene — MgO.SiO₂	0.102				0.102													0.102	100	10.20	
Hypersthene — FeO.SiO₂	0.031			0.031														0.031	132	4.09	
Olivine — 2MgO.SiO₂																			140		O =
Olivine — 2FeO.SiO₂																			204		
Magnetite — FeO.Fe₂O₃			0.013	0.013														0.013	232	3.02	M = 5.00
Hematite — Fe₂O₃			0.016															0.016	160		
Ilmenite — FeO.TiO₂				0.013							0.013							0.013	152	1.98	
Pyrite — FeS₂																			120		
Apatite — 3CaO.P₂O₅.⅓CaF₂						0.007						0.002						0.002	336	0.67	A = 0.97
Calcite — CaO.CO₂						0.003				0.003								0.003	100	0.30	

Salic group = 72.25   Femic group = 27.86

Fig. 4.3: Worked out example — 1

Sep 20. Total allocated silica is 0.162 (step 12) + 0.348 (step 13) + 0.134 (step 15) + 0.034 + 0.026 + 0.008 + 0.102 +0.031(step 19) = 0.845 which is less than available silica 0.982. So the balance SiO_2 (0.982 – 0.845) = **0.137 is to be set as quartz.**

Next step is the calculation of percentages of normative minerals, which is obtained by multiplying the respective molecular proportions (Mol. prop.) by molecular weights (Mol. wt.). These are given below.

Mineral	Mol. prop.	Mol. wt.	% (Norm)
Quartz (Q)	0.137	60	8.22
Orthoclase (Or)	0.027	556	15.01
Albite (Ab)	0.058	524	30.39
Anorthite (An)	0.067	278	18.63
Ca-diopside (Di)	0.034	116	3.94
Mg-diopside (Di)	0.026	100	2.60
Fe-diopside (Di)	0.008	132	1.06
Mg-hypersthene (Hy)	0.102	100	10.20
Fe-hypersthene (Hy)	0.031	132	4.09
Magnetite	0.013	232	3.02
Ilmenite (Il)	0.013	152	1.98
Apatite (Ap)	0.002	336	0.67
Calcite (Cc)	0.003	100	0.30
		Total	**100.11**

Salic group

Q = quartz = 8.22; F = orthoclase + albite + anorthite = 15.01 + 30.39 + 18.63 = 64.03;

L = leucite + nepheline + kaliophilite = 0; C = corundum = 0;

Z = zircon + halite + thenardite = 0

Total salic = Q + F + L + C + Z = 8.22 + 64.03 + 0 + 0 + 0 = 72.25

Femic group

P = acmite + diopside + wollastonite + hypersthene + sodium metasilicate
+ potassium metasilicate

= 0 + 3.94 + 2.60 + 1.06 + 0 + 10.20 + 4.09 + 0 + 0 = 21.89

O = olivine + dicalcium silicate = 0 + 0 = 0

H = hematite + magnetite + chromite = 0 + 3.02 + 0 = 3.02

T = ilmenite + sphene + rutile + perovskite = 1.98 + 0 + 0 + 0 = 1.98

M = H + T = 3.02 + 1.98 = 5

A = apatite + fluorite + pyrite + calcite + cancrinite = 0.67 + 0 + 0 + 0.30 + 0 = 0.97

Total femic = P + O + M + A = 21.89 + 0 + 5 + 0.97 = 27.86

a. Class is Dosalic as salic/femic ratio is 72.25/27.86 = 2.59

b. Order is Perfelic as F (64.03) >> Q (8.22) and L (0)

c. Rang is Alkalicalcic as ($Na_2O + K_2O$) (6.1) = CaO (6.2)

d. Subrang is Dosodic as K_2O (2.5) < Na_2O (3.6)

Calculation of Niggli values and other parameters

$$\text{Total iron (FeO}_t) = 2 \times Fe_2O_3 + FeO = 2 \times 0.013 + 0.063 = 0.089$$

$$Al_2O_3 + Cr_2O_3 = 0.152 + 0 = 0.152$$

$$FeO_t + MgO + MnO = 0.089 + 0.128 + 0.002 = 0.219$$

$$CaO = 0.111$$

$$Na_2O + K_2O = 0.058 + 0.027 = 0.085$$

$$(Al_2O_3 + Cr_2O_3) + (FeO_t + MgO + MnO) + CaO + (Na_2O + K_2O)$$

$$= 0.152 + 0.219 + 0.111 + 0.085 = 0.567$$

i. $$al = \frac{(Al_2O_3 + Cr_2O_3)}{(Al_2O_3 + Cr_2O_3 + 2 \times Fe_2O_3 + FeO + MgO + MnO + CaO + Na_2O + K_2O)} \times 100$$

$$= (0.152 \div 0.567) \times 100 = 26.81$$

ii. $$fm = \frac{(2 \times Fe_2O_3 + FeO + MgO + MnO)}{(Al_2O_3 + Cr_2O_3 + 2 \times Fe_2O_3 + FeO + MgO + MnO + CaO + Na_2O + K_2O)} \times 100$$

$$= (0.219 \div 0.567) \times 100 = 38.62$$

iii. $$c = \frac{CaO}{(Al_2O_3 + Cr_2O_3 + 2 \times Fe_2O_3 + FeO + MgO + MnO + CaO + Na_2O + K_2O)} \times 100$$

$$= (0.111 \div 0.567) \times 100 = 19.58$$

iv. $$alk = \frac{(Na_2O + K_2O)}{(Al_2O_3 + Cr_2O_3 + 2 \times Fe_2O_3 + FeO + MgO + MnO + CaO + Na_2O + K_2O)} \times 100$$

$$= (0.085 \div 0.567) \times 100 = 14.99$$

v. $$si = \frac{SiO_2}{(Al_2O_3 + Cr_2O_3 + 2 \times Fe_2O_3 + FeO + MgO + MnO + CaO + Na_2O + K_2O)} \times 100$$

$$= (0.982 \div 0.567) \times 100 = 173.19$$

vi. $$qz = si - (100 + 4\,alk) = 173.19 - [100 + (4 \times 14.99)] = 13.23$$

vii. $$k = \frac{K_2O}{(Na_2O + K_2O)} \times 100 = \frac{0.027}{(0.058 + 0.027)} \times 100 = 40.26$$

viii. $$mg = \frac{MgO}{(2 \times Fe_2O_3 + FeO + MgO + MnO)} \times 100 = \frac{0.128}{0.219} \times 100 = 58.45$$

ix. Differentiation index (DI) = Q + Or + Ab + Ne + Lc + Ks
$$= 8.22 + 15.01 + 30.39 + 0 + 0 + 0 = 53.62$$

x. Fractionation index (FI) = Q + Or + Ab + Ac + Ns
$$= 8.22 + 15.01 + 30.39 + 0 + 0 = 53.62$$

xi. Crystallisation index (CI) = Anorthite + (2.157003 × Mg-diopside) + (0.700837 × Mg-hypersthene) + Mg-olivine = 18.63 + (2.157003 × 2.6) + (0.700837 × 10.20) + 0 = 31.39

xii. Colour index = Di + Hy + Ol + Mt + Il + Ac
$$= 7.6 + 14.29 + 0 + 3.02 + 1.98 + 0 = 26.89$$

xiii. Agpaitic index $= \dfrac{(Na_2O + K_2O)}{Al_2O_3} = \dfrac{0.058 + 0.027}{0.152} = 0.559$

xiv. Magnesium number $(M') = \dfrac{MgO}{(MgO + FeO)} \times 100 = \dfrac{0.128}{(0.128 + 0.063)} \times 100 = 67.02$

xv. Solidification index $(SI) = \dfrac{MgO}{(MgO + FeO + Fe_2O_3 + Na_2O + K_2O)} \times 100$

$$= \dfrac{5.1}{(5.1 + 4.5 + 2.1 + 3.6 + 2.5)} \times 100 = 28.65$$

xvi. Felsic index $= \dfrac{(Na_2O + K_2O)}{(CaO + Na_2O + K_2O)} \times 100 = \dfrac{(3.6 + 2.5)}{(6.2 + 3.6 + 2.5)} \times 100 = 49.59$

xvii. Mafic index $= \dfrac{(FeO + Fe_2O_3)}{(MgO + FeO + Fe_2O_3)} \times 100 = \dfrac{(4.5 + 2.1)}{(5.1 + 4.5 + 2.1)} \times 100 = 56.41$

xviii. Oxidation ratio $= \dfrac{2Fe_2O_3}{(2FeO + Fe_2O_3)} \times 100 = \dfrac{(2 \times 2.1)}{[(2 \times 2.1) + 4.5]} = 48.28$

xix. The rock plots in the segment VI of Niggli's tetrahedron.

4.6 NORM CALCULATION OF AN UNDERSATURATED ROCK

The chemical analysis of an igneous rock is given in Table 4.4. Calculate the normative mineral composition and determine the Niggli values and other parameters.

TABLE 4.4: Oxide composition of an unsaturated igneous rock

Oxide	Wt.%	Oxide	Wt.%	Oxide	Wt.%	Oxide	Wt.%
SiO_2	34.98	FeO	12.70	Na_2O	1.60	P_2O_5	1.18
Al_2O_3	8.83	MgO	11.67	K_2O	3.30		
Fe_2O_3	4.68	CaO	16.00	TiO_2	4.88		

The step-by-step procedure of calculation of norm is given below and the completed norm form is shown in Fig. 4.4.

Step 1. The percentages are to be converted to molecular proportions (Mol. Prop.) by dividing the weight percentages by respective molecular weights.

Oxide	Mol. Prop.	Oxide	Mol. Prop.
SiO_2	$34.98 \div 60 = 0.583$	CaO	$16.00 \div 56 = 0.286$
Al_2O_3	$8.83 \div 102 = 0.087$	Na_2O	$1.6 \div 62 = 0.026$
Fe_2O_3	$4.68 \div 160 = 0.029$	K_2O	$3.3 \div 94 = 0.035$
FeO	$12.70 \div 72 = 0.176$	TiO_2	$4.88 \div 80 = 0.061$
MgO	$11.67 \div 40 = 0.292$	P_2O_5	$1.18 \div 142 = 0.008$

Constituents of rock	SiO₂	Al₂O₃	Fe₂O₃	FeO	MgO	CaO	Na₂O	K₂O	H₂O	CO₂	TiO₂	P₂O₅	SO₃	S	Cl	F	MnO	Molecular proportions	Molecular weights	Percentages (norm)	Group of standard minerals
Percentages (analysis)	34.98	8.83	4.68	12.7	11.67	16.00	1.60	3.30			4.88	1.18									
Molecular weights	60	102	160	72	40	56	62	94	18	44	80	142	80	64	71	38	71				
Molecular proportions	0.583	0.087	0.029	0.176	0.292	0.286	0.026	0.035			0.061	0.008									
Quartz — SiO₂																			60		Q=
Orthoclase — K₂O.Al₂O₃.6SiO₂	~~0.210~~ ~~0.035~~	~~0.035~~						~~0.035~~										~~0.035~~	556		
Albite — Na₂O.Al₂O₃.6SiO₂	~~0.156~~ ~~0.026~~	~~0.026~~					~~0.026~~											~~0.026~~	524	7.23	F= 7.23
Anorthite — CaO.Al₂O₃.2SiO₂	0.052 0.026	0.026				0.026												0.026	278	7.23	
Leucite — K₂O.Al₂O₃.4SiO₂	0.140 0.035	0.035						0.035										0.035	436	15.26	L= 22.64
Nepheline — Na₂O.Al₂O₃.2SiO₂	0.052 0.026	0.026					0.026											0.026	284	7.38	
Corundum — Al₂O₃		0.099																	102		C=
Larnite — 2CaO.SiO₂	0.099					0.199												0.099	172	17.03	
Acmite — Na₂O.Fe₂O₃.4SiO₂	0.034 0.233																		462		
Diopside — CaO.SiO₂	0.026					0.034 0.233												0.034	116	3.94	P= 7.60
Diopside — MgO.SiO₂	0.179				0.026 0.179													0.026	100	2.60	
Diopside — FeO.SiO₂	0.008 0.054			0.008 0.054														0.008	132	1.06	
Wollastonite — CaO.SiO₂																			116		
Hypersthene — MgO.SiO₂	~~0.113~~				~~0.113~~														100		
Hypersthene — FeO.SiO₂	~~0.032~~			~~0.032~~															132		
Olivine — 2MgO.SiO₂	0.077 0.057				0.153 0.113														140	18.76	O= 43.75
Olivine — 2FeO.SiO₂	0.023 0.016			0.046 0.032															204	7.96	
Magnetite — FeO.Fe₂O₃			0.029	0.029														0.029	232	6.73	
Hematite — Fe₂O₃																			160		M= 16.0
Ilmenite — FeO.TiO₂				0.061							0.061							0.061	152	9.27	
Pyrite — FeS₂				0.061															120		
Apatite — 3CaO.P₂O₅.⅓CaF₂						0.027						0.008				0.005		0.008	336	2.69	A= 2.69
Calcite — CaO.CO₂																			100		

Salic group = 29.87 Femic group = 70.04

Fig. 4.4: Worked out example – 2

Steps 2&3 are not applicable as MnO, NiO, BaO, SrO and ZrO_2 are absent.

Step 4. Set apatite = P_2O_5 (0.008) and allocate $3.33 \times 0.008 = 0.027$ CaO. So **apatite = 0.008**, and balance CaO = $0.286 - 0.027 = 0.259$.

Steps 5–10 are not applicable.

Step 11. FeO = 0.176 and $TiO_2 = 0.061$. Set ilmenite = TiO_2 (0.061) and allocate FeO = 0.061 to it. So **ilmenite = 0.061**. $TiO_2 = 0$ and balance FeO = $0.176 - 0.061 = 0.115$.

Step 12. $Al_2O_3 = 0.087$ and $K_2O = 0.035$. Set orthoclase = K_2O (0.035) and allocate $Al_2O_3 = 0.035$ and $SiO_2 = 6 \times 0.035 = 0.210$ to it. So orthoclase = 0.035, $K_2O = 0$ and balance Al_2O_3 is $0.087 - 0.035 = 0.052$.

Step 13. Available Al_2O_3 is 0.052 and $Na_2O = 0.026$. Set albite = Na_2O (0.026) and allocate 0.026 Al_2O_3 and $6 \times 0.026 = 0.156$ SiO_2 to it. So albite = 0.026, $Al_2O_3 = 0.052 - 0.026 = 0.026$ and $Na_2O = 0$.

Step 14 is not applicable as $Na_2O = 0$.

Step 15. Balance Al_2O_3 is 0.026 and CaO is 0.286. Set anorthite = Al_2O_3 (0.026) and allocate 0.026 CaO and $2 \times 0.026 = 0.052$ SiO_2 to it. So **anorthite = 0.026**, Al_2O_3 is exhausted and CaO balance is $0.259 - 0.026 = 0.233$.

Step 16 is not applicable as $TiO_2 = 0$.

Step 17. FeO = 0.115 (step 11) and $Fe_2O_3 = 0.029$. Set magnetite = Fe_2O_3 (0.029) and allocate 0.029 of FeO. So **magnetite = 0.029**, Fe_2O_3 are exhausted and FeO balance is $0.115 - 0.029 = 0.086$.

Step 18. MgO = 0.292 and FeO = 0.086; a = $[0.292 \div (0.292 + 0.086)] = 0.77$ and b = $[0.0.086 \div (0.292 + 0.086)] = 0.23$.

Step 19. Balance CaO = 0.233 and MgO + FeO = 0.378. Set diopside = 0.233 CaO, $0.77 \times 0.233 = 0.179$ MgO, $0.23 \times 0.233 = 0.054$ FeO and allocate $0.233 + 0.179 + 0.054 = 0.466$ SiO_2 to it. Set remaining MgO $(0.292 - 0.179 = 0.113)$ and FeO $(0.086 - 0.054 = 0.032$ as hypersthene and allocate $0.113 + 0.032 = 0.145$ SiO_2 to it. So Ca-diopside = 0.233, Mg-diopside = 0.179, Fe-diopside = 0.054, Mg-hypersthene = 0.113 and Fe-hypersthene = 0.032. CaO, MgO and FeO are exhausted.

Step 20. Total allocated silica is 0.210 (step 12) + 0.156 (step 13) + 0.052 (step 15) + 0.466 + 0.145 (step 19) = 1.029 which is more than available silica 0.583. So the deficit (D) is $1.029 - 0.583 = 0.446$. This deficit is to be adjusted.

Step 21. Since (D) 0.446 > Hy/2 (0.073), Hy is to be made zero and allocate 0.113 MgO and 0.032 FeO to Mg- and Fe-olivines respectively. SiO_2 to be allocated are (0.113/2) 0.057 and (0.032/2) 0.016 respectively. So Mg-olivine is 0.057 and Fe-olivine is 0.016. D_1 = D (0.446) – Hy/2 (0.073) = 0.373.

Step 22 is not applicable as there is no sphene in the norm. $D_2 = 0.373$.

Step 23. D_2 (0.373) > 4 albite (0.104), so make albite = 0; allocate 0.026 Na_2O, 0.026 Al_2O_3 and $2 \times 0.026 = 0.052$ SiO_2 to nepheline. Nepheline = 0.026. $D_3 = D_2$ (0.373) – 4 Ab (4×0.026) = 0.269.

Step 24. D_3 (0.269) > 2 orthoclase (0.070), so make orthoclase = 0; allocate 0.035 K_2O, 0.035 Al_2O_3 and $4 \times 0.035 = 0.140$ SiO_2 to leucite. **Leucite = 0.035.** $D_4 = D_3$ (0.269) – 2 Or (2×0.035) = 0.199.

Step 25 is not applicable as there is no wollastonite in the norm. $D_5 = 0.199$.

Step 26. D_5 (0.199) $<$ Di/2 (0.233); amounts equal to D_5/2 are to be allocated to dicalcium silicate (Cs) and olivine. Because Cs is not in the norm it is to be created by allocating 0.199 CaO and 0.099 (0.199/2) SiO_2 from the Ca-diopside formed in step 19. This makes **dicalcium silicate = 0.099** and **Ca-diopside = 0.034** (0.223 − 0.199) with CaO = SiO_2 = 0.034. The SiO_2, MgO and FeO of the Mg- and Fe-diopsides (step 19) are to be reallocated to olivine. The allocation of total SiO_2, which is 0.100 (0.199/2) is to be made in the ratio a:b (step 18), i.e. 0.077 and 0.023 of SiO_2 are to be added to Mg- and Fe-olivines formed in step 21. This makes Mg-olivine = 0.057 + 0.077 = 0.134 and Fe-olivine = 0.016 + 0.023 = 0.39 with MgO = 0.113 + (2 × 0.077) = 0.266 (0.001 has been adjusted to make SiO_2 balance 0) and FeO = 0.032 + 2 × 0.023 = 0.78 respectively. The MgO (0.153) and FeO (0.046) added to olivine are to be deducted from diopside. This makes MgO of Mg-diopside 0.026 (0.179 − 0.153) and FeO of Fe-diopside 0.008 (0.054 − 0.046) with 0.026 and 0.008 SiO_2 respectively. The recalculated values of **Mg-diopside = 0.026, Fe-diopside = 0.008, Mg-olivine = 0.134** and **Fe-olivine = 0.039.** This step involved subtraction of 0.466 SiO_2 and addition of 0.267 SiO_2, thus compensating the deficit of 0.199. The calculation of norm is now complete.

Next step is the calculation of percentage of normative minerals, which is obtained by multiplying the respective molecular proportions (Mol. prop.) by molecular weights (Mol. wt.). It is given below:

Mineral	Mol. prop.	Mol. wt.	% (Norm)
Anorthite (An)	0.026	278	7.23
Leucite (Lc)	0.035	436	15.26
Nepheline (Ne)	0.026	284	7.38
Ca-diopside (Di)	0.034	116	3.94
Mg-diopside (Di)	0.026	100	2.60
Fe-diopside (Di)	0.008	132	1.06
Mg-olivine (Ol)	0.134	140	18.76
Fe-olivine (Ol)	0.039	204	7.96
Larnite (Cs)	0.099	172	17.03
Magnetite (Mt)	0.029	232	6.73
Ilmenite (Il)	0.061	152	9.27
Apatite (Ap)	0.008	336	2.69
		Total	**99.91**

Salic group

Q = 0, F = orthoclase + albite + anorthite = 0 + 0 + 7.23 = 7.23
L = leucite + nepheline + kaliophilite = 15.26 + 7.38 + 0 = 22.64
C = corundum = 0, Z = zircon + halite + thenardite = 0
Total salic = Q + F + L + C + Z = 0 + 7.23 + 22.64 + 0 + 0 = 29.87

Femic group

P = acmite + diopside + wollastonite + hypersthene + sodium metasilicate
+ potassium metasilicate
= 0 + 3.94 + 2.60 + 1.06 + 0 + 0 + 0 = 7.60

O = olivine + dicalcium silicate (larnite) = $18.76 + 7.96 + 17.03 = 43.75$

H = hematite + magnetite + chromite = $0 + 6.73 + 0 = 6.73$

T = ilmenite + sphene + rutile + perovskite = $9.27 + 0 + 0 + 0 = 9.27$

$M = H + T = 6.73 + 9.27 = 16.0$;

A = apatite + fluorite + pyrite + calcite + cancrinite = $2.69 + 0 + 0 + 0 + 0 = 2.69$

Total femic = $P + O + M + A = 7.60 + 43.75 + 16.00 + 2.69 = 70.04$

a. Class is Dofemic as salic/femic ratio is $29.87/70.04 = 0.426$

b. Order is Perpolic as $P + O$ (51.35) $>> L$ (22.64)

c. Rang is Domiric as $MgO + FeO + Na_2O$ (25.97) $> CaO$ (16.0)

d. Subrang is Doferrous as MgO (11.67) $< FeO + Na_2O$ (14.3)

Calculation of Niggli values and other parameters

Total iron (FeO_t) = $2 \times Fe_2O_3 + FeO = 2 \times 0.029 + 0.176 = 0.234$

$Al_2O_3 + Cr_2O_3 = 0.087$; $FeO_t + MgO + MnO = 0.234 + 0.292 + 0 = 0.526$, $CaO = 0.286$

$Na_2O + K_2O = 0.026 + 0.035 = 0.061$

$(Al_2O_3 + Cr_2O_3) + (FeO_t + MgO + MnO) + CaO + (Na_2O + K_2O)$
$$= 0.087 + 0.526 + 0.286 + 0.061 = 0.96$$

i. $al = \dfrac{(Al_2O_3 + Cr_2O_3)}{(Al_2O_3 + Cr_2O_3 + 2 \times Fe_2O_3 + FeO + MgO + MnO + CaO + Na_2O + K_2O)} \times 100$

$= (0.087/0.96) \times 100 = 9.06$

ii. $fm = \dfrac{(2 \times Fe_2O_3 + FeO + MgO + MnO)}{(Al_2O_3 + Cr_2O_3 + 2 \times Fe_2O_3 + FeO + MgO + MnO + CaO + Na_2O + K_2O)} \times 100$

$= (0.526/0.96) \times 100 = 54.79$

iii. $c = \dfrac{CaO}{(Al_2O_3 + Cr_2O_3 + 2 \times Fe_2O_3 + FeO + MgO + MnO + CaO + Na_2O + K_2O)} \times 100$

$= (0.286/0.96) \times 100 = 29.79$

iv. $alk = \dfrac{(Na_2O + K_2O)}{(Al_2O_3 + Cr_2O_3 + 2 \times Fe_2O_3 + FeO + MgO + MnO + CaO + Na_2O + K_2O)} \times 100$

$= (0.061/0.96) \times 100 = 6.35$

v. $si = \dfrac{SiO_2}{(Al_2O_3 + Cr_2O_3 + 2 \times Fe_2O_3 + FeO + MgO + MnO + CaO + Na_2O + K_2O)} \times 100$

$= (0.583/0.96) \times 100 = 60.73$

vi. $qz = si - (100 + 4\,alk) = 60.73 - [100 + (4 \times 6.35)] = -64.67$

vii. $k = \dfrac{K_2O}{(Na_2O + K_2O)} \times 100 = \dfrac{0.035}{(0.026 + 0.035)} \times 100 = 57.38$

viii. $mg = \dfrac{MgO}{(2 \times Fe_2O_3 + FeO + MgO + MnO)} \times 100 = \dfrac{0.292}{0.526} \times 100 = 55.51$

ix. Differentiation index (DI) = Q + Or + Ab + Ne + Lc + Ks = 0 + 0 + 0 + 7.38 + 15.26 + 0 = 22.64

x. Fractionation index (FI) = Q + Or + Ab + Ac + Ns = 0 + 0 + 0 + 0 + 0 = 0

xi. Crystallisation index (CI) = Anorthite + (2.157003 × Mg-diopside) + (0.700837 × Mg-hypersthene) + Mg-olivine = 7.23 + (2.157003 × 2.6) + (0.700837 × 0) + 18.76 = 31.60

xii. Colour index = Di + Hy + Ol + Mt + Il + Ac = 7.60 + 0 + 26.72 + 6.73 + 9.27 + 0 = 50.32

xiii. Agpaitic index = $\dfrac{(Na_2O + K_2O)}{Al_2O_3} = \dfrac{0.026 + 0.035}{0.087} = 0.70$

xiv. Magnesium number (M′) = $\dfrac{MgO}{(MgO + FeO)} \times 100 = \dfrac{0.292}{(0.292 + 0.176)} \times 100 = 62.39$

xv. Solidification index (SI) = $\dfrac{MgO}{(MgO + FeO + Fe_2O_3 + Na_2O + K_2O)} \times 100$

$$= \dfrac{11.67}{(11.67 + 12.70 + 4.68 + 1.60 + 3.30)} \times 100 = 34.37$$

xvi. Felsic index = $\dfrac{(Na_2O + K_2O)}{(CaO + Na_2O + K_2O)} \times 100 = \dfrac{(1.6 + 3.3)}{(16.0 + 1.6 + 3.3)} \times 100 = 23.44$

xvii. Mafic index = $\dfrac{(FeO + Fe_2O_3)}{(MgO + FeO + Fe_2O_3)} \times 100 = \dfrac{(12.70 + 4.68)}{(11.67 + 12.70 + 4.68)} \times 100 = 59.83$

xviii. Oxidation ratio = $\dfrac{2Fe_2O_3}{(2FeO + Fe_2O_3)} \times 100 = \dfrac{(2 \times 4.68)}{[(2 \times 4.68) + 12.70]} \times 100 = 42.43$

xix. The rock plots in the segment IV of Niggli's tetrahedron.

4.7 PROBLEMS FOR NORM CALCULATION

The oxide compositions of some igneous rocks are given in Table 4.5. Determine the normative composition and calculate the Niggli values in each case.

TABLE 4.5: Oxide compositions of igneous rocks

Rock	SiO_2	TiO_2	Al_2O_3	Fe_2O_3	FeO	MnO	MgO	CaO	Na_2O	K_2O	P_2O_5	CO_2
1	73.22	0.32	12.65	1.95	1.45	0.10	0.20	0.50	4.60	4.90	0.10	0.01
2	73.12	0.31	13.28	1.65	1.35	0.21	0.53	1.55	3.45	4.38	0.08	0.09
3	72.65	0.32	13.95	1.55	1.08	0.20	0.50	1.55	3.75	4.35	0.10	0.00
4	72.25	0.45	10.35	3.45	2.45	0.15	0.45	0.52	5.25	4.56	0.10	0.02
5	72.15	0.25	13.15	1.60	1.80	0.15	0.65	1.55	3.55	4.85	0.25	0.05
6	71.63	0.31	14.36	1.25	1.68	0.05	0.71	1.87	3.85	4.12	0.12	0.05
7	66.95	0.35	16.65	2.55	1.45	0.24	1.55	3.34	4.16	2.58	0.15	0.03
8	66.55	0.37	15.78	1.52	2.78	0.25	1.83	3.95	3.85	2.85	0.25	0.02
9	66.18	0.45	15.67	2.14	2.94	0.20	1.85	3.86	3.57	2.82	0.30	0.02
10	62.25	0.55	17.95	2.25	2.34	0.16	1.30	2.50	5.50	5.10	0.10	0.00

(Contd...)

TABLE 4.5: Oxide compositions of igneous rocks (*Contd...*)

Rock	SiO_2	TiO_2	Al_2O_3	Fe_2O_3	FeO	MnO	MgO	CaO	Na_2O	K_2O	P_2O_5	CO_2
11	62.15	0.45	17.58	2.94	1.64	0.16	0.85	1.46	6.68	5.95	0.12	0.02
12	61.75	0.55	16.48	1.83	3.82	0.12	2.85	5.42	3.65	3.13	0.25	0.15
13	60.78	0.47	17.75	2.78	2.88	0.15	1.16	3.25	4.56	5.88	0.30	0.04
14	59.75	0.45	18.29	3.32	4.35	0.08	0.52	1.95	6.11	5.02	0.16	0.00
15	59.55	0.55	16.58	2.75	4.65	0.10	3.96	6.34	3.28	1.91	0.30	0.03
16	59.38	0.70	16.68	3.56	3.64	0.10	3.60	6.54	3.65	1.95	0.20	0.00
17	58.65	0.95	16.45	3.76	3.24	0.10	3.25	4.62	3.64	4.96	0.38	0.00
18	58.24	0.87	17.02	3.37	4.44	0.14	3.43	6.89	3.62	1.72	0.21	0.05
19	57.68	0.40	20.75	2.84	1.46	0.15	0.65	1.64	8.85	5.48	0.10	0.00
20	57.65	1.02	16.87	2.64	5.02	0.12	3.84	6.75	3.84	1.85	0.30	0.10
21	56.87	0.80	16.84	3.45	4.64	0.10	4.58	6.87	3.41	2.10	0.30	0.04
22	56.25	1.02	17.68	3.84	3.46	0.20	2.68	5.46	5.31	3.60	0.50	0.00
23	55.36	0.65	16.64	3.32	4.45	0.10	4.26	7.20	3.51	4.12	0.39	0.00
24	54.99	0.62	21.65	2.58	2.46	0.15	0.96	2.42	8.23	5.58	0.16	0.20
25	54.65	0.85	19.95	3.64	2.43	0.30	1.02	2.68	8.45	5.64	0.30	0.09
26	53.84	1.82	16.82	2.45	7.36	0.20	5.02	7.35	3.24	1.70	0.20	0.00
27	51.35	1.80	16.59	3.54	7.86	0.17	3.25	10.12	2.85	2.20	0.22	0.05
28	50.44	1.02	16.64	2.46	7.49	0.14	8.85	9.64	2.28	0.71	0.15	0.18
29	50.40	0.10	28.30	1.25	1.35	0.12	1.42	12.5	3.7	0.76	0.1	0.00
30	50.28	0.75	25.56	1.46	2.58	0.05	2.19	12.48	3.15	1.24	0.12	0.14
31	50.25	1.12	15.88	3.35	7.72	0.12	7.68	9.84	2.49	1.24	0.24	0.07
32	50.05	1.55	13.37	3.71	10.39	0.25	6.49	11.16	2.38	0.45	0.15	0.05
33	49.20	1.84	15.86	3.89	7.48	0.20	6.95	9.68	3.19	1.25	0.35	0.11
34	48.85	1.45	15.85	5.68	6.82	0.30	6.45	8.94	3.45	1.71	0.50	0.00
35	48.68	1.90	18.64	4.56	5.76	0.20	4.26	8.90	4.30	2.30	0.50	0.00
36	48.26	1.03	18.24	3.45	6.48	0.10	7.75	11.09	2.50	0.90	0.20	0.00
37	47.56	2.50	15.80	3.65	7.96	0.16	7.17	10.10	3.20	1.40	0.50	0.00
38	46.00	0.40	4.40	2.40	14.40	0.10	26.2	5.30	0.50	0.30	0.00	0.00
39	45.89	0.08	1.60	0.00	6.92	0.11	43.46	1.16	0.26	0.22	0.30	0.00
40	45.75	0.35	2.35	1.84	7.52	0.05	35.6	4.65	0.62	0.67	0.40	0.20
41	45.60	1.70	8.40	2.40	10.20	0.10	21.7	7.60	1.40	0.40	0.30	0.20
42	45.35	1.70	16.60	5.12	7.64	0.10	5.46	9.47	5.45	2.51	0.60	0.00
43	44.60	0.90	11.00	1.00	12.50	0.20	14.10	13.10	1.10	1.50	0.00	0.00
44	44.20	0.13	2.05	0.05	8.29	0.13	42.21	1.92	0.27	0.30	0.40	0.05
45	43.95	1.57	3.88	0.75	7.50	0.13	38.20	2.60	0.60	0.22	0.40	0.20
46	43.68	2.19	9.83	2.25	8.65	0.01	15.24	15.82	1.65	0.68	0.00	0.00
47	42.80	1.60	18.90	3.90	4.88	0.20	3.54	10.64	9.68	2.46	1.40	0.00
48	42.26	0.63	5.24	3.61	7.66	0.41	31.00	6.15	1.52	1.12	0.10	0.30
49	41.26	1.20	4.98	5.62	8.35	0.10	32.20	4.51	0.58	1.10	0.10	0.02
50	39.26	2.30	12.60	6.40	8.40	0.10	11.60	12.90	3.86	1.78	0.80	0.00
51	69.92	0.49	13.91	1.18	3.01	0.10	0.45	1.73	4.27	4.92	0.02	0.00
52	66.12	1.03	15.91	2.73	3.25	0.31	1.00	1.68	3.07	4.68	0.22	0.00
53	54.08	2.08	9.49	3.19	3.34	0.00	6.74	3.55	4.42	11.76	1.35	0.00
54	53.81	0.95	17.79	2.44	6.60	0.19	5.87	8.79	2.76	0.62	0.18	0.00
55	53.80	0.30	18.72	4.99	3.59	0.00	1.78	2.80	8.82	5.20	0.00	0.00

(*Contd...*)

TABLE 4.5: Oxide compositions of igneous rocks (*Contd...*)

Rock	SiO_2	TiO_2	Al_2O_3	Fe_2O_3	FeO	MnO	MgO	CaO	Na_2O	K_2O	P_2O_5	CO_2
56	49.96	0.80	22.30	2.54	2.25	0.00	0.99	3.86	8.11	6.04	1.27	1.88
57	47.16	3.36	14.45	1.61	13.81	0.00	5.24	8.13	3.95	2.20	0.05	0.04
58	46.42	0.18	15.92	2.44	5.65	0.09	11.50	16.14	1.44	0.17	0.05	0.00
59	46.06	0.73	10.01	3.17	5.61	0.04	14.60	10.55	4.02	5.14	0.05	0.02
60	45.82	7.28	7.86	6.07	4.98	0.02	10.9	4.70	1.70	8.82	1.85	0.00
61	44.40	1.53	19.95	5.15	3.98	0.04	1.75	8.49	6.50	8.14	0.05	0.02
62	54.56	0.53	15.85	0.95	6.07	0.17	8.71	8.89	3.05	1.18	0.04	0.00
63	68.32	0.31	15.26	1.66	2.50	0.04	1.47	3.24	4.27	2.81	0.12	0.00
64	48.08	1.17	17.22	1.32	8.44	0.16	8.62	11.38	2.37	1.02	0.22	0.00
65	45.08	0.94	16.42	2.29	9.29	0.06	11.65	10.46	2.06	1.57	0.18	0.00
66	48.15	2.64	18.02	2.52	9.50	0.12	5.25	10.17	3.46	0.14	0.03	0.00
67	60.23	1.18	11.29	5.52	9.11	0.24	0.51	5.11	3.92	2.62	0.27	0.00
68	50.60	1.28	14.80	0.00	9.67	0.00	9.05	11.72	2.51	0.15	0.22	0.00
69	46.70	3.60	17.30	3.80	7.10	0.00	4.70	9.70	4.10	3.00	0.00	0.00
70	48.50	2.68	17.53	4.43	4.75	0.12	4.52	6.32	6.79	3.74	0.62	0.00
71	43.56	2.26	7.85	5.57	5.86	0.15	11.43	11.89	3.74	7.19	0.50	0.00
72	35.52	3.97	6.50	7.23	7.52	0.24	14.08	16.79	3.32	4.09	0.74	0.00
73	46.16	2.45	15.32	3.50	9.67	0.21	7.25	9.58	3.87	1.34	0.65	0.00
74	47.10	2.30	14.70	3.50	9.00	0.20	8.70	10.20	3.10	0.90	0.30	0.00
75	49.58	3.17	13.19	2.40	9.49	0.12	8.30	10.69	2.25	0.55	0.26	0.00

4.8 ANSWERS

The answers are given in Table 4.6

TABLE 4.6: Answers of problems given in Table 4.5
(*The calculated values may differ slightly depending on the number of
significant digits taken after the decimal point in different formula weights*)

Mineral	1	2	3	4	5	6	7	8	9	10	11	12	13
Quartz	25.85	31.41	29.34	27.57	28.09	27.78	22.26	21.08	22.42	2.54	0.00	12.66	2.48
Albite	37.79	29.19	31.73	27.85	30.04	32.58	35.20	32.58	30.21	46.54	49.78	30.89	38.59
Anorthite	0.00	6.60	7.04	0.00	5.62	8.18	15.40	17.36	17.06	9.23	0.41	19.34	10.60
Orthoclase	28.96	25.88	25.71	26.95	28.66	24.35	15.25	16.84	16.67	30.14	35.16	18.50	34.75
Nepheline	0.00	0.00	0.00	0.00	0.00	0.00	0.00	0.00	0.00	0.00	3.65	0.00	0.00
Corundum	0.00	0.45	0.49	0.00	0.00	0.57	1.37	0.00	0.49	0.00	0.00	0.00	0.00
Diopside	1.54	0.00	0.00	1.57	0.10	0.00	0.00	0.39	0.00	2.10	5.00	4.06	2.79
Hypersthene	2.87	4.05	3.58	6.98	4.71	4.17	7.51	8.40	9.16	6.00	0.00	9.96	6.43
Olivine	0.00	0.00	0.00	0.00	0.00	0.00	0.00	0.00	0.00	0.00	2.59	0.00	0.00
Acmite	1.00	0.00	0.00	5.35	0.00	0.00	0.00	0.00	0.00	0.00	0.00	0.00	0.00
Na_2SiO_3	0.00	0.00	0.00	2.44	0.00	0.00	0.00	0.00	0.00	0.00	0.00	0.00	0.00
Ilmenite	0.61	0.59	0.61	0.85	0.47	0.59	0.66	0.70	0.85	1.04	0.85	1.04	0.89
Magnetite	1.05	1.38	1.20	0.00	1.57	1.36	1.81	2.00	2.35	2.12	2.07	2.64	2.60
Apatite	0.23	0.19	0.23	0.23	0.58	0.28	0.35	0.58	0.70	0.23	0.28	0.58	0.70
Calcite	0.02	0.20	0.00	0.05	0.11	0.11	0.07	0.05	0.05	0.00	0.05	0.34	0.09
Total	99.92	99.94	99.93	99.84	99.95	99.96	99.88	99.98	99.96	99.94	99.84	100.0	99.91

(Contd...)

TABLE 4.6: Answers of problems given in Table 4.5 (*Contd...*)
(*The calculated values may differ slightly depending on the number of significant digits taken after the decimal point in different formula weights*)

Niggli values

al	39.98	41.24	42.62	30.34	39.21	41.71	39.66	36.44	36.00	38.34	38.09	32.35	36.71
fm	16.42	17.61	15.49	27.06	19.27	17.01	22.87	25.18	27.33	20.78	18.35	29.81	22.33
c	2.88	8.77	8.63	2.78	8.42	9.89	14.49	16.61	16.15	9.73	5.76	19.38	12.24
alk	40.72	32.39	33.27	39.82	33.10	31.38	22.97	21.77	20.52	31.15	37.80	18.46	28.71
si	393.4	386.0	377.3	360.0	365.7	353.7	271.1	261.3	258.4	226.1	228.9	206.1	213.7
qz	130.5	156.5	144.3	100.8	133.3	128.2	79.2	74.2	76.4	1.5	-22.3	32.3	-1.1

Mineral	14	15	16	17	18	19	20	21	22	23	24	25	26
Quartz	0.00	12.61	10.84	3.82	10.15	0.00	8.06	6.96	0.00	0.00	0.00	0.00	4.52
Albite	50.47	27.75	30.89	30.80	30.63	34.32	32.49	28.85	42.95	29.70	26.79	24.51	27.42
Anorthite	7.65	24.88	23.37	13.90	25.11	0.71	23.33	24.44	13.77	17.48	5.65	0.00	26.33
Orthoclase	29.67	11.29	11.52	29.31	10.17	32.38	10.93	12.41	21.27	24.35	32.98	33.33	10.05
Nepheline	0.67	0.00	0.00	0.00	0.00	21.98	0.00	0.00	1.07	0.00	23.21	25.30	0.00
Diopside	0.85	3.61	6.34	5.34	6.05	5.76	6.17	6.04	8.27	12.79	3.41	9.11	7.17
Hypersthene	0.00	14.57	11.81	10.80	11.95	0.00	12.57	15.15	0.00	6.83	0.00	0.00	16.05
Olivine	5.84	0.00	0.00	0.00	0.00	1.77	0.00	0.00	6.07	3.04	3.55	2.18	0.00
Acmite	0.00	0.00	0.00	0.00	0.00	0.00	0.00	0.00	0.00	0.00	0.00	0.25	0.00
Ilmenite	0.85	1.04	1.33	1.80	1.65	0.76	1.94	1.52	1.94	1.23	1.18	1.61	3.46
Magnetite	3.55	3.45	3.31	3.20	3.61	1.94	3.58	3.74	3.35	3.60	2.31	2.63	4.63
Apatite	0.37	0.70	0.46	0.88	0.49	0.23	0.70	0.70	1.16	0.90	0.37	0.70	0.46
Calcite	0.00	0.07	0.00	0.00	0.11	0.00	0.23	0.09	0.00	0.00	0.45	0.20	0.00r
Total	99.92	99.97	99.86	99.85	99.92	99.85	100.0	99.90	99.86	99.92	99.90	99.82	100.1

Niggli values

al	37.19	29.64	29.93	30.43	29.90	40.05	29.12	28.05	30.92	27.04	39.30	35.65	26.27
fm	24.07	36.36	34.13	32.96	34.32	14.60	35.29	37.97	29.59	35.01	17.13	19.86	41.61
c	7.22	20.65	21.37	15.57	22.06	5.77	21.22	20.84	17.39	21.31	8.00	8.72	20.91
alk	31.52	13.35	14.57	21.04	13.74	39.58	14.37	13.14	22.11	16.65	35.57	35.77	11.21
si	206.55	181.0	181.1	184.5	173.9	189.3	169.2	161.0	167.2	152.9	169.7	166.0	142.8
qz	-19.5	27.6	22.8	0.32	18.96	-69.1	11.7	8.5	-21.2	-13.7	-72.6	-77.1	-1.9

Mineral	27	28	29	30	31	32	33	34	35	36	37	38	39
Quartz	1.56	0.44	0.00	0.00	0.00	2.49	0.00	0.00	0.00	0.00	0.00	0.00	0.00
Albite	24.12	19.29	27.27	26.26	21.07	20.14	26.99	27.76	23.59	21.15	24.24	4.24	2.2
Anorthite	25.98	33.07	58.37	51.94	28.49	24.47	25.26	22.71	24.77	35.89	24.61	8.88	2.55
Orthoclase	13.00	4.20	4.49	7.33	7.33	2.66	7.39	10.11	13.59	5.32	8.27	1.77	1.30
Nepheline	0.00	0.00	2.19	0.21	0.00	0.00	0.00	0.78	6.93	0.00	1.54	0.00	0.00
Diopside	18.63	10.16	2.45	6.73	14.81	24.24	16.00	14.98	13.17	14.36	17.91	13.87	0.98
Hypersthene	7.35	25.53	0.00	0.00	17.97	16.06	5.23	0.00	0.00	4.90	0.00	32.42	27.23
Olivine	0.00	0.00	3.57	3.62	2.32	0.00	9.25	13.82	8.29	11.29	12.09	30.38	61.77
Ilmenite	3.42	1.94	0.19	1.42	2.13	2.94	3.49	2.75	3.61	1.96	4.75	0.76	0.15
Magnetite	5.34	4.70	1.20	1.88	5.19	6.64	5.31	5.77	4.77	4.64	5.44	8.00	3.35
Apatite	0.51	0.35	0.23	0.28	0.56	0.35	0.81	1.16	1.16	0.46	1.16	0.00	0.70
Calcite	0.11	0.41	0.00	0.32	0.16	0.11	0.25	0.00	0.00	0.00	0.00	0.00	0.00
Total	100.0	100.09	99.96	99.99	100.0	100.1	99.99	99.84	99.87	99.97	100.01	100.3	100.2

(*Contd...*)

TABLE 4.6: Answers of problems given in Table 4.5 (*Contd...*)
(*The calculated values may differ slightly depending on the number of significant digits taken after the decimal point in different formula weights*)

Niggli values

al	25.03	22.12	43.35	38.73	21.40	17.96	21.53	21.58	26.80	23.68	20.95	4.17	1.28
fm	36.48	48.54	11.18	16.93	47.12	48.83	45.57	46.00	36.13	43.48	45.67	85.61	96.50
c	27.81	23.34	34.88	34.45	24.15	27.30	23.94	22.17	23.31	26.23	24.39	9.14	1.69
alk	10.68	6.01	10.59	9.89	7.33	5.91	8.97	10.25	13.76	6.61	8.99	1.09	0.53
si	131.7	114.0	131.3	129.5	115.1	114.3	113.6	113.1	119.0	106.5	107.2	74.0	62.33
qz	−11.0	−10.1	−11.1	−10.0	−14.2	−9.4	−22.3	−28.0	−36.1	−19.9	−28.8	−30.3	−39.8

Mineral	40	41	42	43	44	45	46	47	48	49	50	51
Quartz	0.00	0.00	0.00	0.00	0.00	0.00	0.00	0.00	0.00	0.00	0.00	21.25
Albite	5.25	11.85	7.61	0.23	2.28	5.08	0.00	0.00	0.00	3.01	0.00	36.13
Anorthite	1.65	15.45	13.42	20.65	3.50	7.24	17.41	0.85	4.17	7.74	11.80	4.26
Orthoclase	3.96	2.36	14.83	8.86	1.77	1.30	0.00	0.22	6.46	6.50	0.00	29.08
Nepheline	0.00	0.00	20.86	4.92	0.00	0.00	7.56	44.37	6.97	1.03	17.69	0.00
Leucite	0.00	0.00	0.00	0.00	0.00	0.00	3.15	11.23	0.13	0.00	8.25	0.00
Diopside	13.79	15.07	24.26	35.58	2.43	1.39	42.71	30.67	18.78	11.04	22.27	3.60
Hypersthene	15.88	16.19	0.00	0.00	19.00	20.66	0.00	0.00	0.00	0.00	0.00	2.76
Wollastonite	0.00	0.00	0.00	0.00	0.00	0.00	0.00	2.21	0.00	0.00	0.00	0.00
Olivine	53.11	28.89	8.37	21.92	65.97	56.20	17.75	0.00	56.13	61.57	20.80	0.00
Larnite	0.00	0.00	0.00	0.00	0.00	0.00	2.23	0.00	0.00	0.00	5.96	0.00
Ilmenite	0.66	3.23	3.23	1.71	0.25	2.98	4.16	3.04	1.20	2.28	4.37	0.93
Magnetite	4.44	5.97	5.92	6.48	4.03	3.96	5.16	4.06	5.28	6.48	6.84	1.97
Apatite	0.93	0.70	1.39	0.00	0.93	0.93	0.00	3.24	0.23	0.23	1.85	0.05
Calcite	0.45	0.45	0.00	0.00	0.11	0.45	0.00	0.00	0.68	0.00	0.00	0.00
Total	100.1	100.2	99.89	100.4	100.3	100.2	100.1	99.89	100.0	99.88	99.83	100.0

Niggli values

al	2.02	8.57	21.57	11.76	1.63	3.26	10.23	24.21	4.55	4.27	13.38	38.13
fm	89.21	74.50	40.83	59.05	94.98	91.73	56.19	27.15	82.51	86.84	52.87	19.35
c	7.28	14.13	22.41	25.51	2.78	3.98	29.99	24.82	9.72	7.05	24.95	8.64
alk	1.50	2.79	15.19	3.68	0.61	1.03	3.59	23.82	3.22	1.84	8.80	33.89
si	66.8	79.1	100.2	81.1	59.7	62.8	77.3	93.2	62.3	60.2	70.9	325.8
qz	−39.2	−32.1	−60.6	−33.6	−42.8	−41.3	−37.1	−102.1	−50.6	−47.2	−64.3	90.2

Mineral	52	53	54	55	56	57	58	59	60	61	62	63
Quartz	23.77	0.00	6.56	0.00	0.00	0.00	0.00	0.00	0.00	0.00	1.29	22.53
Albite	25.98	0.00	23.34	15.57	22.01	22.72	6.24	0.00	0.00	0.00	25.81	36.13
Anorthite	6.90	0.00	34.32	0.00	0.00	15.20	36.47	0.00	0.00	1.22	26.07	14.17
Orthoclase	27.66	51.81	3.66	30.73	35.69	13.00	1.00	0.00	39.99	0.00	6.97	16.61
Nepheline	0.00	0.00	0.00	28.05	24.20	5.80	3.22	12.39	0.00	29.80	0.00	0.00
Leucite	0.00	0.00	0.00	0.00	0.00	0.00	0.00	23.82	2.30	37.72	0.00	0.00
Corundum	3.27	0.00	0.00	0.00	2.80	0.00	0.00	0.00	0.00	0.00	0.00	0.00
Diopside	0.00	7.01	6.58	11.79	0.00	20.13	34.52	16.77	8.79	13.16	14.20	0.92
Hypersthene	7.12	10.87	19.11	0.00	0.00	0.00	0.00	0.00	0.00	0.00	21.35	6.81

(*Contd...*)

TABLE 4.6: Answers of problems given in Table 4.5 (*Contd...*)
(*The calculated values may differ slightly depending on the number of significant digits taken after the decimal point in different formula weights*)

Olivine	0.00	5.10	0.00	5.96	4.24	9.53	14.30	27.28	17.25	2.98	0.00	0.00
Larnite	0.00	0.00	0.00	0.00	0.00	0.00	0.00	9.55	0.00	7.70	0.00	0.00
Acmite	0.00	5.99	0.00	6.43	0.00	0.00	0.00	8.16	10.07	0.00	0.00	0.00
K_2SiO_3	0.00	4.90	0.00	0.00	0.00	0.00	0.00	0.00	2.55	0.00	0.00	0.00
Na_2SiO_3	0.00	7.12	0.00	0.00	0.00	0.00	0.00	0.44	0.69	0.00	0.00	0.00
Ilmenite	1.96	3.95	1.80	0.57	1.52	6.38	0.34	1.39	13.83	2.91	1.01	0.59
Magnetite	2.75	0.00	4.25	0.68	2.19	7.38	3.80	0.00	0.00	4.16	3.35	1.93
Apatite	0.51	3.13	0.42	0.00	2.94	0.12	0.12	0.12	4.29	0.12	0.09	0.28
Calcite	0.00	0.00	0.00	0.00	3.90	0.09	0.00	0.05	0.00	0.00	0.00	0.00
Na_2CO_3	0.00	0.00	0.00	0.00	0.39	0.00	0.00	0.00	0.00	0.00	0.00	0.00
Total	99.91	99.88	100.1	99.78	99.88	100.4	100.0	99.97	99.76	99.77	100.1	99.97

Niggli values

al	39.60	15.31	26.67	31.22	38.33	19.76	18.00	11.04	11.01	27.84	22.45	37.49
fm	27.58	41.93	41.52	26.66	15.38	47.84	45.87	54.33	59.68	23.34	45.70	23.25
c	7.62	10.43	24.00	8.51	12.08	20.25	33.24	21.19	11.99	21.58	22.93	14.50
alk	25.21	32.32	7.81	33.61	34.20	12.15	2.89	13.44	17.32	27.25	8.92	24.75
si	279.8	148.4	137.1	152.5	146.0	109.6	89.2	86.3	109.1	105.3	131.4	285.4
qz	78.9	–81.0	5.9	–81.9	–90.8	–39.0	–22.3	–67.4	–60.2	–103.7	–4.31	86.36

Mineral	*64*	*65*	*66*	*67*	*68*	*69*	*70*	*71*	*72*	*73*	*74*	*75*
Quartz	0.00	0.00	0.00	14.58	0.01	0.00	0.00	0.00	0.00	0.00	0.00	2.61
Albite	20.05	9.47	29.28	33.17	21.24	10.82	15.43	0.00	0.00	21.34	23.67	19.04
Anorthite	33.34	30.92	33.22	5.47	28.67	19.94	6.31	0.00	0.00	20.47	23.54	24.27
Orthoclase	6.03	9.28	0.83	15.48	0.89	17.73	22.10	0.00	0.00	7.92	5.32	3.25
Nepheline	0.00	4.31	0.00	0.00	0.00	12.93	22.77	0.19	5.78	6.18	1.39	0.00
Leucite	0.00	0.00	0.00	0.00	0.00	0.00	0.00	33.32	0.00	0.00	0.00	0.00
Kalsilite	0.00	0.00	0.00	0.00	0.00	0.00	0.00	0.00	13.73	0.00	0.00	0.00
Diopside	17.43	15.88	13.86	15.92	22.47	22.68	16.91	1.98	0.00	18.48	20.22	21.68
Hypersthene	2.50	0.00	7.12	5.63	19.43	0.00	0.00	0.00	0.00	0.00	0.00	17.04
Olivine	13.46	22.61	5.07	0.00	0.00	3.94	5.59	27.39	33.78	13.33	14.99	0.00
Larnite	0.00	0.00	0.00	0.00	0.00	0.00	0.00	16.49	24.29	0.00	0.00	0.00
Acmite	0.00	0.00	0.00	0.00	0.00	0.00	0.00	10.47	13.54	0.00	0.00	0.00
Na_2SiO_3	0.00	0.00	0.00	0.00	0.00	0.00	0.00	4.52	0.48	0.00	0.00	0.00
Ilmenite	2.22	1.79	5.01	2.24	2.43	6.84	5.09	4.29	7.54	4.65	4.37	6.02
Magnetite	4.65	5.48	5.68	6.80	4.67	5.09	4.22	0.00	0.00	6.19	5.87	5.63
Apatite	0.51	0.42	0.07	0.63	0.51	0.00	1.44	1.16	1.71	1.51	0.70	0.60
Total	100.2	100.2	100.1	99.92	100.3	99.97	99.86	99.81	100.9	100.1	100.1	100.13

Niggli values

al	21.85	19.00	24.81	21.93	19.17	24.08	25.65	8.90	6.30	19.76	18.60	17.63
fm	45.49	53.07	41.63	41.94	47.63	37.42	35.23	50.74	54.43	47.65	50.21	50.59
c	26.30	22.04	25.51	18.08	27.65	24.59	16.84	24.55	29.66	22.50	23.51	26.03
alk	6.35	5.89	8.05	18.05	5.56	13.92	22.28	15.82	9.60	10.09	7.69	5.75
si	103.7	88.66	112.7	198.9	111.4	110.5	120.6	88.9	58.6	101.2	101.3	112.7
qz	–21.7	–34.9	–19.5	26.7	–10.8	–45.2	–68.5	–79.3	–79.8	–39.1	–29.5	–10.3

5

Modal Analysis

While *norm* refers to the hypothetical mineral content of an igneous rock, mode indicates the percentage (by weight) of the individual minerals, which make up a rock of any type. Determination of mode is carried out by any one of the following methods.

  i. *Direct determination*: In this process the rock is crushed to liberate the minerals and volume of each constituent mineral is determined.

 ii. *Area measurement method*: In this method the area of each mineral exposed on a flat surface (polished or thin section) is found out, which is proportional to the volume of each mineral.

iii. *Delesse and Rosiwal method*: In this method a series of closely spaced linear traverses are made across the thin section and total length of the line intercepted by constituent minerals is found out. Total intercept length of a particular mineral is proportional to the area and hence to the volume.

 iv. *Point counting method*: This method is a variant of Delesse and Rosiwal method, in which the intercept length of a particular mineral is measured in a series of discrete steps instead of continuity. The steps are counted and the numbers of counts (points) for all the minerals are recorded. The numbers of points are proportional to the linear intercept, and thus with the area and volume of the minerals present.

In laboratory, modal analysis is commonly done by point counting method. The counting can be made either manually or by an automatic point counter. The steps involved in modal analysis are given below.

Step 1. Examine the rockslide under petrological microscope and identify the constituent minerals. The eyepiece of the microscope in which counting is to be made should have cross wires.

Step 2. Fit the slide holding attachment on the stage of the microscope securely with the help of screws. Check the lateral movement of the slide holder. It should move freely without any hindrance. Keep the slide within the holder.

Step 3. Attach the point counter with the slide holding attachment. Fix the number of points the slide to move in one stroke. Normally it is kept at '1'. However, if the minerals are few in number and are of bigger in size, the counter may be set at 2 or more. Assign different keys to different minerals and note them.

Step 4. Set the initial reading of the counter at '0'. The upper limit may be fixed if required; otherwise it is set at '9999'. When the numbers of points counted reach the upper limit, the buzzer will give the indication and the counter will not move further. It is necessary in case of counting of fixed number of heavy minerals in sedimentary rocks.

Step 5. Bring the top left corner of the slide under the cross wires. Note the mineral and strike the corresponding key. With one stroke, the cross wires will appear to move the number of points set at step 3 from left to right. In fact, the cross wire is stationary and the slide moves from right to left. Continue the process till the cross wires reach the end of the line.

Step 6. Rotate the knob of the slide holder to select the second line and drag it till the left hand side of the slide comes under cross wires. Repeat the counting process till end of that line. Continue the selection of line and counting processes till the bottom right corner of the slide comes under the cross wires.

Step 7. Determine the percent of each constituent mineral from point count data.

5.1 MODAL ANALYSIS OF AN IGNEOUS ROCK

A thin section of an igneous rock is shown in Fig. 5.1. The percentages of constituent minerals are to be determined by modal analysis.

The rock is composed of four types of minerals, viz. quartz, potash feldspar (orthoclase and microcline), plagioclase and pyroxene. A total number of 12 linear traverses have been taken and the positions of cross wires (indicated by cross) at each striking position are shown. The point count data are given in Table 5.1. Out of the total number of 128 counts, 50 are on quartz, 52 are on potash feldspar, 24 on plagioclase and 2 are on pyroxene. The analysis shows that the rock is composed of $(50 \div 128) \times 100 = 39.06\%$ quartz, $(52 \div 128) \times 100 = 40.63\%$ potash feldspar, $(24 \div 128) \times 100 = 18.75\%$ plagioclase and $(2 \div 128) \times 100 = 1.56\%$ pyroxene. The rock is granite. All the possible views of the rock slide are to be examined.

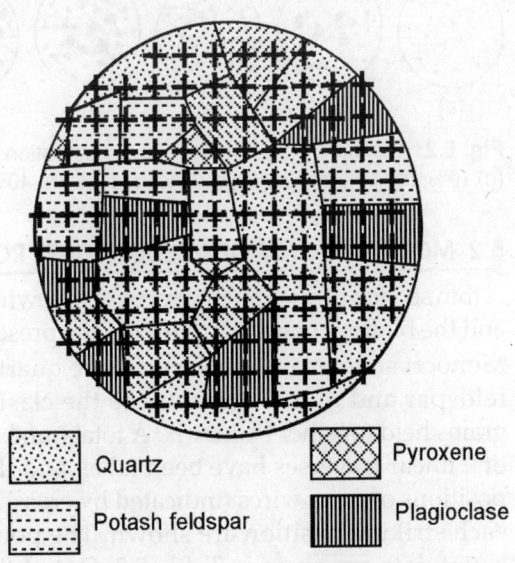

Quartz

Potash feldspar

Pyroxene

Plagioclase

Fig. 5.1: Thin section of an igneous rock

In some instances, it may be necessary to quickly determine the percentages of different constituent minerals. In such cases, percentages of area occupied by different minerals are determined by visual estimation by comparing with a set of standard diagrams (Figs 5.2 a to i) proposed by Terry and Chilingar. Since the common rocks are composed of a number of minerals, suitable staining is necessary to distinguish and differentiate a particular mineral from others.

TABLE 5.1: Point count data of igneous rock section shown in Fig. 5.1

Line	Quartz	K-feldspar	Plagioclase	Pyroxene	Total
1	4	2	0	0	6
2	7	2	1	0	10
3	2	6	2	1	11
4	2	9	0	1	12
5	3	9	0	0	12
6	3	4	6	0	13
7	5	4	4	0	13
8	9	2	1	0	12
9	6	5	1	0	12
10	2	6	3	0	11
11	3	3	4	0	10
12	4	0	2	0	6
Total	50	52	24	2	128
Percent	39.06	40.63	18.75	1.56	100

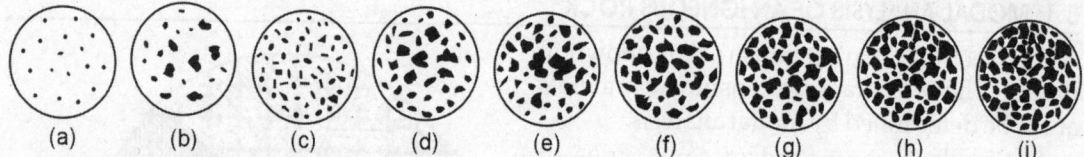

(a) (b) (c) (d) (e) (f) (g) (h) (i)

Fig. 5.2: Standard diagrams for visual estimation of modal proportion of minerals in rocks: (a) 1%, (b) 5%, (c) 10%, (d) 15%, (e) 20%, (f) 25%, (g) 30%, (h) 40%, (i) 50%

5.2 MODAL ANALYSIS OF A SEDIMENTARY ROCK

A thin section of a sedimentary rock is shown in Fig. 5.3. The percentages of constituent minerals and the binding material (matrix in the present case) are to be determined by modal analysis. Monocrystalline and polycrystalline quartz, feldspar and rock fragments are the clastic grains held together by matrix. A total number of 9 linear traverses have been taken and the positions of cross wires (indicated by cross) at each striking position are shown. The point count data are given in Table 5.2. Out of the total number of 135 counts, 34 are monocrystalline quartz, 36 are polycrystalline quartz, 25 are feldspar, 18 are rock fragment and 22 are matrix. The analysis shows that the rock is composed of 51.9% quartz, 18.5% feldspar, 13.3% rock fragment and 16.3% matrix. From modal analysis, the rock is identified to be feldspathic greywacke or arkosic wacke.

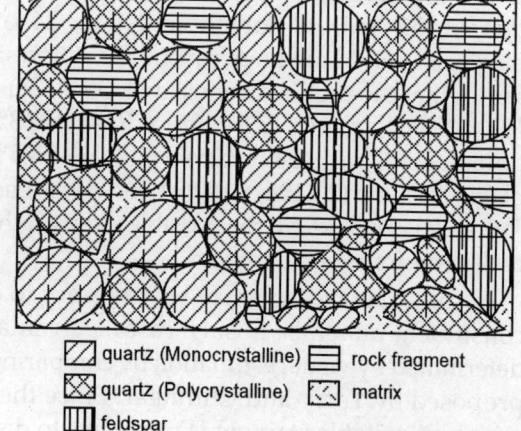

quartz (Monocrystalline)   rock fragment
quartz (Polycrystalline)   matrix
feldspar

Fig. 5.3: Thin section of a sandstone

TABLE 5.2: Modal analysis data of sedimentary rock shown in Fig.5.3

Line	Monocrystalline quartz	Polycrystalline quartz	Feldspar	Rock fragment	Matrix and cement	Total
1	4	4	2	4	1	15
2	6	1	2	3	3	15
3	6	3	2	2	2	15
4	1	5	4	0	5	15
5	2	4	4	2	3	15
6	3	3	4	3	2	15
7	2	6	2	4	1	15
8	5	3	3	0	4	15
9	5	7	2	0	1	15
Total	34	36	25	18	22	135
percent	25.2	26.7	18.5	13.3	16.3	100

6

Graphic Plot of Petrochemistry Data

The data generated by chemical analysis of rocks are used in various ways. Computation of norm from these data has already been discussed. Graphic plot of petrochemistry in variation, triangular and discriminant diagrams are discussed below.

6.1 VARIATION DIAGRAM

A variation diagram is commonly a graph in which the chemical characters of a series of rocks are plotted against each other (or against a dominant constituent like SiO_2). The variation of chemical characters can also be plotted with reference to certain physical properties like colour index, specific gravity, etc. or vice versa. The commonest type is the Harker diagram, in which percentages of elements (as oxides) are plotted against silica percentages. The chemical analysis data of common igneous rocks like gabbro, quartz-gabbro, diorite, quartz-diorite, granodiorite and granite in terms of some major oxides are given in Table 6.1 and line-graphs of different oxides with respect to SiO_2 are plotted in Fig. 6.1. The line-graphs indicate variation of constituent oxides with respect to SiO_2 and thus are variation diagrams. They show characteristic trends. Al_2O_3, CaO, FeO and MgO percentages decrease while those of Na_2O and K_2O increase with increase of silica content of different rocks. It is known that with progressive differentiation of basic magma, gabbro, quartz-gabbro, diorite, quartz-diorite, granodiorite and granite are produced in succession. The silica percentage, which indicates the basic or acidic

TABLE 6.1: Chemical compositions (major oxides) of some common igneous rocks

Rock	Gabbro	Quartz-gabbro	Diorite	Quartz-diorite	Granodiorite	Granite
SiO_2	48.3	53.9	57.2	59.9	66.6	71.5
Al_2O_3	18.1	17.4	16.9	16.6	15.3	13.3
CaO	11.0	8.6	7.2	6.2	3.8	1.3
FeO	6.1	5.2	4.7	4.1	2.2	2.0
MgO	7.6	5.5	4.4	3.8	2.1	0.7
Na_2O	2.6	3.2	3.3	3.4	2.5	3.8
K_2O	0.9	1.9	2.5	2.5	3.5	4.9

nature of a rock, shows marked increase with progressive differentiation of basic magma resulting in the formation of more acidic rocks. From the variation diagram it is apparent that amounts of SiO_2, Na_2O and K_2O increase, whereas amounts of Al_2O_3, CaO, FeO and MgO decrease with progressive differentiation of basic magma that leads to the formation of acidic rocks. The line-diagrams shown in Fig. 6.1 reveal genetic relationships between a series of rocks. Physical properties like specific gravity, refractive index etc of many rocks and minerals are controlled by their chemical compositions. Figure 6.2 shows the variation of all the three refractive indices of plagioclase feldspars. The variation diagram suggests that all the three refractive indices increase from sodic (albite) to calcic (anorthite) end-members, i.e. refractive indices increase with decrease of sodium and increase of calcium contents.

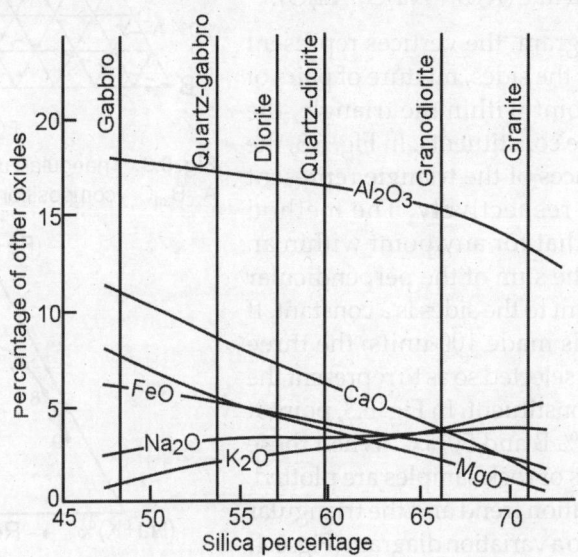

Fig. 6.1: Plot of different oxides with respect to SiO_2 in different igneous rocks

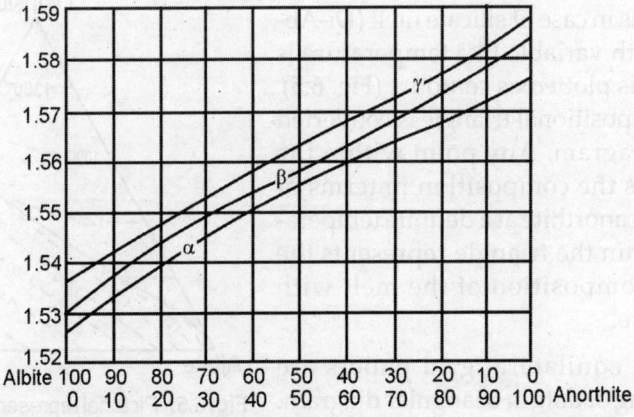

Fig. 6.2: Graph showing variation of refractive indices of plagioclase feldspars with chemical composition

6.2 TRIANGULAR DIAGRAM

Triangular diagrams are extensively used in petrology to display the composition of a rock in terms of three selected components. These three components may comprise the whole system or a part of it. For example, the composition of a sediment can be expressed in terms of components like sand, silt and clay; where sand + silt + clay = 100. Sometimes one or more of the component(s) may be composite consisting of more than one constituent as in case of total alkali like ($K_2O + Na_2O + Li_2O$).

In a triangular diagram, the vertices represent 100% of a constituent; the sides, mixture of pairs of constituents and a point within the triangle, the mixture of all the three constituents. In Fig. 6.3, the top, left and right apices of the triangle represent 100% of A, B and C respectively. The method depends on the fact that for any point within an equilateral triangle, the sum of the perpendicular distances from the point to the sides is a constant. If this constant length is made 100 units, the three perpendiculars can be selected so as to represent the percentages of each constituent. In Fig. 6.3, point X represents 20% A, 30% B and 50% C. When these constituents of a series of rock samples are plotted, they may show a variation trend and the triangular diagram is converted to a variation diagram (Fig. 6.4) in which the rock composition changes from 1 to 9 in the direction of the arrow. The sequence of change of composition may be due to differentiation. In some instances, as in case of silicate melt (Di-Ab-An system), a fourth variable like temperature is introduced, which is plotted as contours (Fig. 6.5). As a result, the compositional triangle is converted into a variation diagram. Any point within the triangle represents the composition in terms of diopside, albite and anorthite at a definite temperature. Any line within the triangle represents the variation of the composition of the melt with varying temperature.

Specially ruled equilateral grid papers are available for plotting of data in triangular diagram. As a customary, 5 or 10% intervals are shown by

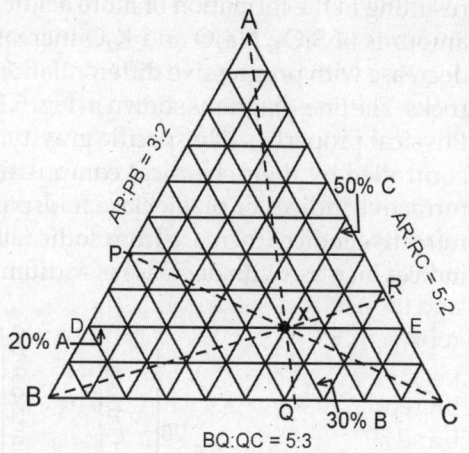

Fig. 6.3: Triangular ruled paper showing plot of $A_{20}B_{30}C_{50}$ composition point

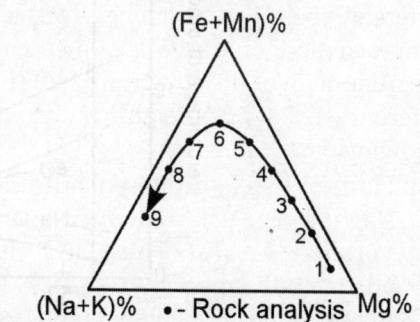

Fig. 6.4: Triangular diagram showing variation of elements in different types of rocks

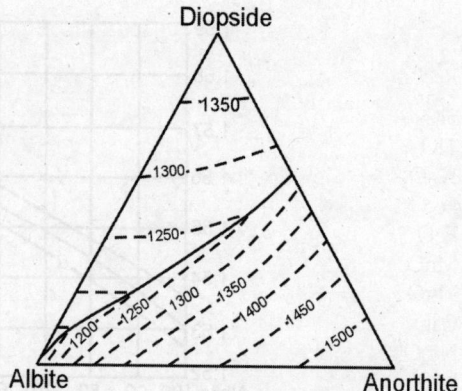

Fig. 6.5: Pictorial representation of dipside-albite-anorthite system with changing temperature

dark or bold lines and 1 or 2% lines are indicted by lighter and faint lines. Such a ruled paper with lines at 10% intervals is shown in Fig. 6.3. The base line BC represents all points with 0% A and point A represents 100% A. Similar is the case for other two sides and corresponding apical points. The second line from BC towards A (DE) indicates 20% A. Similarly the third line from AC and fifth line from AB indicate 30% B and 50% C respectively. The point of intersection of these three lines (X) represents composition of 20% A, 30% B and 50% C, i.e. $A_{20}B_{30}C_{50}$.

In the absence of ruled lines, the point can be plotted by taking the ratios of any two components. The ratio A:B = 20:30 or 2:3. The position of the point P on line AB is decided in such a manner that AP:PB = 3:2. It is obtained by dividing the line AB into 5 (3 + 2) parts. All points on line CP represent A: B = 2:3. The ratio B:C = 30:50 or 3:5. The line BC is divided into 8 (5 + 3) parts and the position of the point Q is fixed in such a manner that BQ:QC = 5:3. All points on line AQ represent B:C = 3: 5. The A:C ratio is equal to 20:50 = 2:5. The line AC is divided into 7 (2 + 5) parts and the position of the point R is fixed in such a manner that AR: RC = 5:2. All points on line BR represent A:C = 2:5. The lines AQ, BR and CP intersect at a point that represents 20% A, 30% B and 50% C, or $A_{20}B_{30}C_{50}$. In actual practice only two lines out of 3 (AQ, BR and CP) are drawn.

6.3 DISCRIMINATION DIAGRAM

Different types of rocks can be effectively discriminated from each other on the basis of their chemical composition. A number of such diagrams have been proposed by different workers. Some of them are illustrated in the following paragraphs.

    i. Harpum (1963) proposed a discriminant diagram (Fig. 6.6) for acid igneous rocks based on Na_2O versus K_2O plot. The Na_2O versus K_2O plots adequately identify the rocks 1, 2, 3 and 4 (Table 6.2) to be granite, adamelite, granodiorite and tonalite respectively and discriminate them from each other (Fig. 6.6).

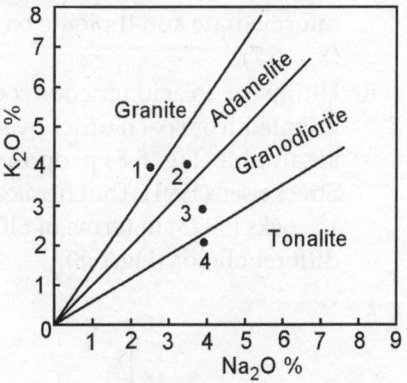

Fig. 6.6: Discriminant diagram of acidic rocks (modified after Harpum, 1963)

TABLE 6.2: Chemical compositions of some common igneous rocks								
Rock	1	2	3	4	5	6	7	8
SiO_2	72.30	68.65	66.04	61.42	50.52	49.20	51.54	51.50
TiO_2	0.32	0.55	0.52	0.54	1.00	2.40	3.53	0.98
Al_2O_3	14.43	14.60	15.64	15.49	14.28	16.64	14.74	15.78
Fe_2O_3	1.21	1.21	1.38	1.63	2.01	3.72	4.92	2.21
FeO	1.74	2.72	2.75	3.82	6.38	6.20	7.34	7.40
MnO	0.05	0.08	0.08	0.07	0.14	0.16	0.18	0.14
MgO	0.72	1.15	1.72	2.80	8.23	5.18	5.56	8.75
CaO	1.94	2.67	3.63	5.41	8.64	7.92	7.91	9.41
Na_2O	2.51	3.50	3.90	3.92	4.60	3.97	3.98	2.27
K_2O	3.99	4.00	2.90	2.11	3.60	2.54	1.63	0.71
P_2O_5	0.12	0.19	0.18	0.24	0.15	0.59	0.00	0.15
CO_2	0.06	0.06	0.07	0.15	0.16	0.10	0.00	0.16

TABLE 6.3: Silica (SiO_2) and total alkali ($Na_2O + K_2O$) contents of some igneous rocks

Rock	SiO_2	$Na_2O + K_2O$	Rock	SiO_2	$Na_2O + K_2O$	Rock	SiO_2	$Na_2O + K_2O$
9	42.14	11.04	14	50.87	10.21	19	60.05	4.27
10	44.85	3.25	15	54.78	4.46	20	61.85	7.25
11	45.58	7.95	16	56.02	8.28	21	68035	4.55
12	50.01	2.63	17	58.8	10.80	22	64.02	8.96
13	50.06	8.32	18	60.05	13.05	23	73.00	6.85

ii. Saggerson and Willam (1964), McDonald and Katsura (1964) and Mohr (1976) proposed a discriminant diagram (Fig. 6.7) to discriminate between different types of basalts on the basis of silica versus total alkali plot. SiO_2 versus ($Na_2O + K_2O$) plots of rocks 5, 6, 7 and 8 (Table 6.2) suggest them to be high-alkaline, sub-alkaline, intermediate and tholeiite basalts respectively (Fig. 6.7).

iii. Ultrabasic to acid igneous rocks can be discriminated from each other by silica versus total alkali plot (Fig. 6.8) proposed by Le Bas and Streckeisen (1991). The chemical compositions of 15 rocks (9–23) in terms of SiO_2 and ($Na_2O + K_2O$) are given in Table 6.3, which plot in different fields (Fig. 6.8).

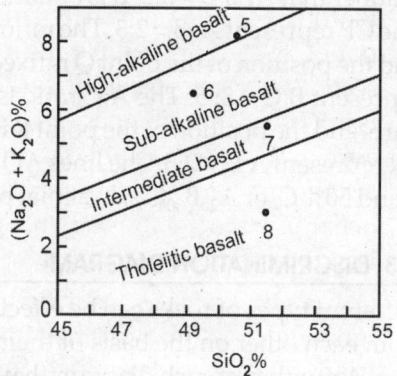

Fig. 6.7: Discriminant diagram of basalts (modified after Mohr, 1976 and others)

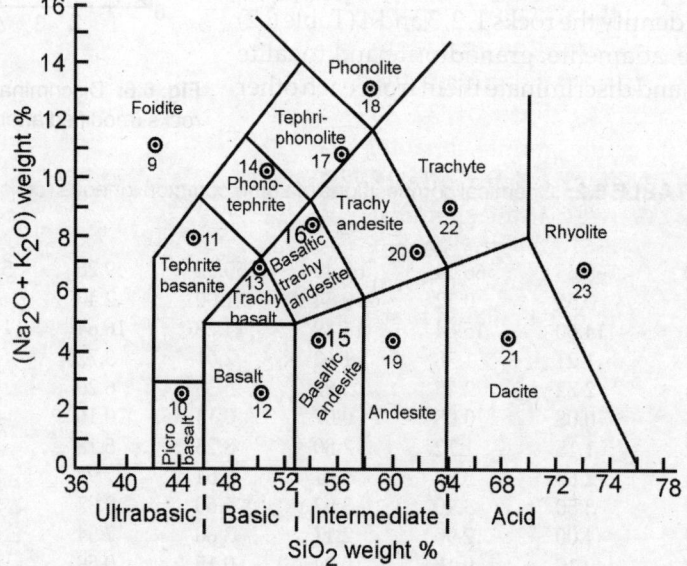

Fig. 6.8: Discriminant diagram of igneous rocks (modified after LeBas and Streckeisen, 1991)

iv. $TiO_2 - (MnO \times 10) - (P_2O_5 \times 10)$ ternary diagram (Fig. 6.9) proposed by Mullen (1983) discriminate between important rocks of basalt family. The TiO_2, MnO and P_2O_5 contents of some igneous rocks are given in Table 6.4. The data cannot be plotted directly. The MnO and P_2O_5 values are to be multiplied by 10 and the sum $TiO_2 + (MnO \times 10) + (P_2O_5 \times 10)$ is found out. TiO_2, $(MnO \times 10)$ and $(P_2O_5 \times 10)$ values are recalculated making the sum 100%. Plotting is made in the triangular ruled paper by the procedure indicated in section 6.2. From the plottings (Fig. 6.9) it is apparent that the rocks 24, 25, 26, 27 and 28 are calc-alkaline basalt, island arc tholeiite, mid-oceanic ridge basalt, oceanic island andesite and oceanic island tholeiite respectively.

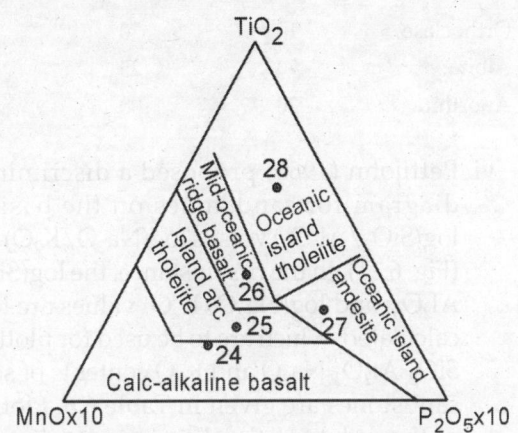

Fig. 6.9: Discriminant diagram of basaltic rocks (modified after Mullen, 1983)

v. O' Corner (1965) discriminated the acid igneous rocks into different types by triangular plot of normative anorthite, albite and orthoclase (Fig. 6.10). The normative orthoclase, albite and anorthite data of six rocks (29–34) are given in Table 6.5. The data are to be recalculated making the sum anorthite + albite + orthoclase equal to 100 and plotted in the triangular ruled paper by the procedure indicated in section 6.2. The results indicate that the rocks 29, 30, 31, 32, 33 and 34 are trondhjemite, granite, tonalite, granodiorite, adamelite and quartz monzonite respectively. If the chemical analysis data in terms of weight percent of oxides are provided instead of norm values, the norm is to be calculated first.

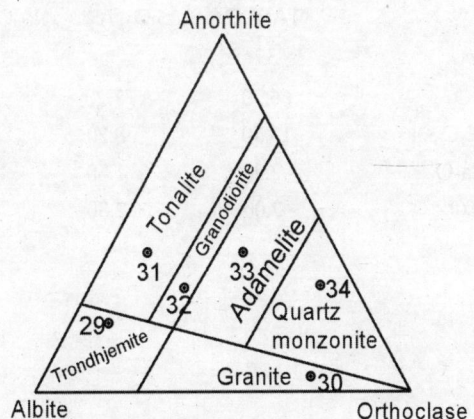

Fig. 6.10: Discriminant diagram of acidic rocks (modified after O' Corner, 1965)

TABLE 6.4: TiO_2, MnO and P_2O_5 contents of some igneous rocks				
Rock	24	25	26	27	28
TiO_2	0.06	0.14	0.17	0.75	0.19
MnO	0.24	0.45	0.17	0.19	0.15
P_2O_5	0.13	0.25	0.16	0.31	0.41

TABLE 6.5: Recalculated values of normative orthoclase, albite and anorthite contents of some igneous rocks

Rock	29	30	31	32	33	34
Orthoclase	10	70	10	25	35	60
Albite	70	25	50	45	25	10
Anorthite	20	05	40	30	40	30

vi. Pettijohn (1963) proposed a discriminant diagram for sandstones on the basis of $\log(SiO_2/Al_2O_3)$ versus $\log(Na_2O/K_2O)$ plot (Fig. 6.11). In the first instance, the $\log(SiO_2/Al_2O_3)$ and $\log(Na_2O/K_2O)$ values are to be calculated, which are to be used for plotting. SiO_2, Al_2O_3, Na_2O and K_2O contents of some sandstones are given in Table 6.6. Plotting of these data in the discriminant diagram (Fig. 6.11) indicate that the sandstones 35, 36, 37, 38, 39 and 40 are greywacke, arkose, lithic arenite, subarkose, sublithic arenite and quartz arenite respectively.

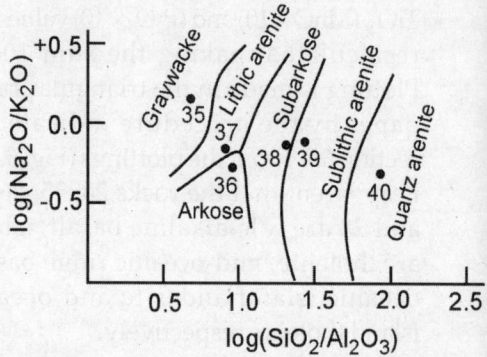

Fig. 6.11: Discriminant diagram of sandstones (modified after Pettijohn, 1963)

TABLE 6.6: SiO_2, Al_2O_3, Na_2O and K_2O contents of some sandstones

Rock	35	36	37	38	39	40
SiO_2	66.70	77.20	66.20	92.13	92.60	95.40
Al_2O_3	13.50	8.70	8.10	4.42	3.50	1.10
Na_2O	2.90	1.50	0.90	0.11	0.54	0.10
K_2O	2.00	2.80	1.30	0.72	0.71	0.20

Part II

Metamorphic Petrology

7. Fabric and Petrography of Metamorphic Rocks

8. Graphic Construction of ACF, AKF and AFM Diagrams

Fabric and Petrography of Metamorphic Rocks

Metamorphism may be defined as the response of solid rocks to pronounced changes in temperature, pressure and chemical environment excluding those of weathering and cementation. The changes in temperature, pressure and chemical environment upset the physical and chemical equilibrium of the rocks, as a result of which changes in texture, structure and mineral assemblage take place, which are more stable in the changed environment. The metamorphic rocks derived from igneous rocks are termed orthometamorphic, while those derived from sedimentary rocks are known as parametamorphic. The common metamorphic minerals are andalusite, chlorite, biotite, garnet, staurolite, kyanite, sillimanite, etc.

Under the physical and chemical conditions imposed during metamorphism, rocks undergo internal adjustment that tends towards a state of structural as well as chemical equilibrium. When the metamorphism is incipient, many of the textural and structural features of the parent rock are preserved. These are termed inherited fabric (*palimset*) and provide valuable clue to decipher the parentage of the changed rock. When the metamorphism is ideally complete, a new assemblage of minerals and a new set of fabric develop, which completely replace the old ones. These are termed *crystalloblastic*.

7.1 SHAPE OF METAMORPHIC MINERALS

Like the minerals of igneous rocks, minerals constituting the metamorphic rocks may show either well-developed or ill-developed grain boundaries. Two types of shapes are pronounced in case of metamorphic rocks. These are

 i. *Idioblastic shape*: The shape of the grain is termed idioblastic, when all the possible crystal faces are developed. Garnet, staurolite, etc. show well-developed idioblastic shape.

 ii. *Xenoblastic shape*: The term xenoblastic is used when the mineral grain has poorly developed outlines. The minerals are anhedral with interlocking boundaries.

7.2 METAMORPHIC TEXTURE

Metamorphic textures can be broadly divided into two types, viz. foliated and nonfoliated. Foliated texture develops due to the parallel arrangement of platy or flaky minerals like biotite,

53

chlorite, serpentine etc. The foliated rocks split parallel to the plane of foliation. Some of the foliated textures are:

7.2.1 Lepidoblastic Texture

This term is used when micaceous minerals are dominant.

7.2.2 Nematoblastic Texture

This term is used when fibrous minerals are dominant.

In case of non-foliated textures, the mineral grains do not show any preferred arrangement. Some of the non-foliated textures are:

7.2.3 Crystalloblastic Texture

It refers to better developed recrystallised minerals, which show euhedral shape with inter-locking grain boundaries.

7.2.4 Granoblastic Texture

In this case the principal constituents are granular and equidimensional.

7.2.5 Porphyroblastic Texture

When large crystals with fairly developed crystal faces are surrounded by fine-grained groundmass the term porphyroblastic is used.

7.2.6 Poikiloblastic Texture

This term is used to sieve-structure in which large crystals like garnet, staurolite etc include a number of mineral inclusions.

7.2.7 Cataclastic Texture

This type of texture develops due to mechanical breaking of rocks.

7.2.8 Hornfelsic Texture

A rock is said to have hornfelsic texture when it is composed of a mosaic of equidimensional and unoriented minerals.

7.3 METAMORPHIC STRUCTURE

Important metamorphic structures are granulose, schistose, gneissose, maculose, cataclastic, phyllitic and slaty.

7.3.1 Granulose Structure

Granulase structure is due to the predominance of equidimensional grains such as quartz, feldspar, calcite, dolomite etc in a metamorphic rock (Fig. 7.1A). The cleavable, lamellar or elongated rod shaped grains are either absent or present in subordinate amount.

7.3.2 Schistose Structure

Schistose structure develops due to the presence of flacky, tabular and highly cleavable minerals like mica, chlorite etc or rod-shaped amphibole in a metamorphic rock (Fig. 7.1B). Under the influence of directed pressure, these minerals are arranged in layers and the rocks develop schistocity, a property that enable the rocks to split along planes.

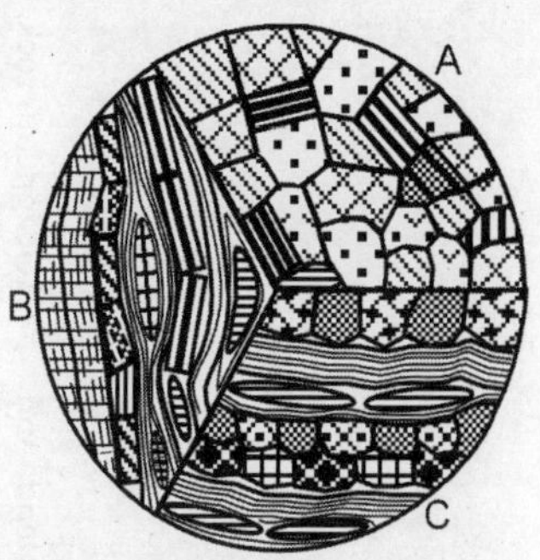

7.3.3 Gneissose Structure

It is a composite structure that develops due to the alternation of granulose and schistose bands and lenticles, which differ in mineral composition and texture (Fig. 7.1C). The micas and hornblends may occur segregated into more or less schistose bands or lenticles within a granulose matrix.

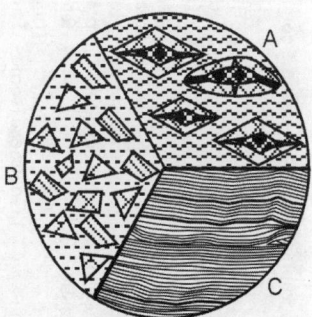

Fig. 7.1: Metamorphic structures: A. Granulose, B. Schistose, C. Gneissose

7.3.4 Maculose Structure

In this type of structure, porphyroblasts of relatively stronger minerals such as andalusite, cordierite, etc. are well developed or in which spotting appears as a result of incipient crystallization of these minerals and segregation of carbonaceous matter (Fig. 7.2A).

7.3.5 Cataclastic Structure

Broken and fragmented rocks produced by shearing are characterized by cataclastic structure (Fig. 7.2B). The softer rock like shale develops cleavage while relatively harder rocks are crushed into powder with formation of crushed breccia. The minerals of cataclastic rocks show strain effects like undulose extinction in quartz and secondary twining in feldspar and calcite.

Fig. 7.2: Metamorphic structures: A. Maculose, B. Cataclastic, C. Phyllitic

7.3.6 Phyllitic Structure

This texture is caused by parallel arrangement of flaky minerals (Fig. 7.2C) and shown by phyllites.

7.3.7 Slaty Structure

This type of texture is caused by parallel arrangement of microscopic minerals (mostly clay minerals) and shown by slate.

7.4 PETROGRAPHY OF METAMORPHIC ROCKS

The petrographic description of some common metamorphic rocks is given in Table 7.1.

TABLE 7.1: Petrographic description of some common metamorphic rocks

Rock	Physical character	Texture	Structure	Essential minerals	Accessory minerals	Parent rock
Acid–charnockite	Greenish black in colour, massive, hard, compact	Medium- to coarse-grained	Granulose	Quartz (green), alkali feldspars, hypersthene, plagioclase	Biotite, hornblende	Acid igneous rock
Amphibolite	Greenish black in colour, massive, hard, compact	Medium- to coarse-grained, granoblastic, nematoblastic	Gneissose (sometimes schistose)	Hornblende, quartz, plagioclase, alkali feldspars (often keolinised)	Biotite	Argillaceous sediment
Augen-gneiss	Alternating dark and light bands, foliated	Coarse- to fine-grained, large grains are lensoidal (eye-shaped), porphyroblastic	Gneissose (sometimes cataclastic)	Large grains of quartz and feldspar (eye-shaped)	Hornblende and other minerals	Acid igneous rock, granite
Basic granulite	Dark coloured with greenish patches	Medium- to coarse-grained, crystalloblastic	Granulose	Quartz, feldspar, pyroxene	Biotite and others	Basic igneous rock
BHJ	Alternate bands of iron-black and reddish brown colours of variable thickness	Very fine-grained	Banded	Hematite (iron black) and jasper (reddish brown)	Nil	—
Calc-schist	Light coloured, soft, reacts with acid	Medium- to coarse-grained, laminated, foliated	Schistose	Calcite, dolomite, quartz	Biotite, muscovite, chlorite	Calcareous shale
Calc-silicate	Variable coloured	Coarse-grained, crystalloblastic, equigranular	Granulose	Calcite, dolomite, wollastonite, garnet, plagioclase, diopside	Quartz, amphibole, sphene, graphite, zoisite, muscovite	Impure limestone
Charnockite	Dark coloured, massive, hard, compact	Medium- to coarse-grained	Granulose	Quartz, feldspar, hypersthene, plagioclase	Pyroxene, hornblende, biotite	Acid igneous rock, granite
Chlorite-schist	Greenish (often shining)	Medium-grained, foliated, crystalloblastic	Schistose (schistose planes are wavy in some cases)	Chlorite	Orthoclase, quartz, epidote, biotite, amphibole, garnet, staurolite	Argillaceous sediment
Eclogite	Light to dark coloured	Granoblastic, non-foliated	Gneissose	Plagioclase, garnet, diopside, jadeite, quartz	Hornblende, kyanite, rutile, corundum, dolomite, aragonite	Basaltic rock

(Contd...)

TABLE 7.1: Petrographic description of some common metamorphic rocks (*Contd...*)

Rock	Physical character	Texture	Structure	Essential minerals	Accessory minerals	Parent rock
Gneiss	Light to dark coloured, banded, foliated	Fine- to coarse-grained, granular, foliated, alternation of bands and lenses, poorly defined schistocity	Gneissose	Quartz, orthoclase, plagioclase, biotite, muscovite, hornblende	Garnet, kyanite, sillimanite	Siliceous sediment, acid igneous rock
Gondite	Variable coloured	Medium- to coarse-grained, granoblastic, foliated	Banded	Manganese minerals, spessartite, quartz	Rhodonite, braunite, cryptomelane	Manganiferous sediment
Granite gneiss	Light to dark coloured, banded, foliated	Fine- to coarse-grained, granular, foliated, alternation of bands and lenses, poorly defined schistocity	Gneissose	Quartz, orthoclase, plagioclase	Biotite, muscovite, hornblende, garnet	Granite
Granulite	Light to dark coloured	Fine- to medium-grained, granoblastic, interlocking	Granulose	Quartz, feldspar, garnet	Pyroxene, amphibole, calcite, kyanite	Granite, sandstone
Greenstone	Light to dark green coloured, hard	Fine- to medium-grained, granoblastic, foliated, oriented grains	Granulose	Chlorite	Epidote, biotite, muscovite, garnet, plagioclase	Basic igneous rock
Hornblende-gneiss	Light and dark bands	Fine-grained, crystalloblastic, granoblastic, nematoblastic	Gneissose	Hornblende, quartz, feldspar	Biotite, pyroxene	Granitic rocks, sandstone etc. with ferromagnesian component
Hornblende-schist	Dark coloured	Coarse-grained, crystalloblastic	Schistose	Hornblende, biotite, feldspar	Epidote, quartz, sphene	Basic lava
Hornfels	Gray, black, green, variably coloured	Fine-grained, angular, porphyroblastic	Maculose, stratified	Quartz, plagioclase, pyroxene, sillimanite, cordierite	Biotite, muscovite, chlorite, hornblende, garnet, epidote	Shale
Hypersthene granulite	Dark coloured, hard, compact	Coarse-grained, crystalloblastic	Granulose	Hypersthene, quartz, orthoclase, plagioclase	Biotite, muscovite	Argillaceous rock
Khondalite	Irregular light to dark coloured bands	Medium- to coarse-grained, crystalloblastic, granoblastic, porphyroblastic	Schistose/ gneissose	Quartz, feldspar, garnet, sillimanite	Biotite, epidote, graphite (occasionally)	Argillaceous rock
Leptynite	White with reddish brown tint	Medium- to coarse-grained, crystalloblastic, granoblastic	Granulose	Quartz, feldspar, garnet	Biotite	Migmatisation and feldspathisation of igneous rock

(*Contd...*)

TABLE 7.1: Petrographic description of some common metamorphic rocks (Contd...)

Rock	Physical character	Texture	Structure	Essential minerals	Accessory minerals	Parent rock
Magnetite-quartzite	Light and dark coloured banding	Fine- to medium-grained, banded	Palimset	Quartz, magnetite	—	Banded hematite quartzite
Marble	Variously coloured, reacts with acid, soft	Fine- to coarse-grained, granoblastic, compact, interlocking	Granulose	Calcite, dolomite	Pyroxene, garnet, wollastonite, olivine, serpentine etc.	Calcitic/dolomitic limestone
Meta-conglomerate	Variously coloured, hard and compact	Granoblastic	Cataclastic	Quartz, feldspar, deformed pebbles, sheared and crushed grains	—	Quartzite, conglomerate, granite
Migmatite	Mixed rock, light (granitic component) and dark (metamorphic rock)	Granoblastic, migmatitic, quartz and feldspars are very coarse-grained	Gneissose	Quartz, feldspar	Amphiboles, pyroxenes, micas	Light coloured igneous rock mixed with dark coloured metamorphic parent rock
Mylonite	Dark coloured	Non-foliated, sheared, crushed, finely granular	Cataclastic	Crushed angular grains	Siliceous and micaceous minerals	Quartzite, conglomerate, granite or any other type of rock
Phyllite	Variously coloured, soft	Foliated, grains ultra-fine-grained	Schistose, foliated, cleaved	Argillaceous materials, muscovite	Chlorite, biotite	Shale, tuff
Pyroxene-marble	Buff coloured with dark gray patches	Fine- to coarse-grained, crystalloblastic	Granulose	Calcite, pyroxene	Dolomite, amphibole	Impure calcareous sediments
Quartzite	White, gray, massive, hard, compact, break with conchoidal fracture, vitreous lustre	Fine- to medium-grained, crystalloblastic	Granulose	Quartz	Feldspar, mica, garnet	Sandstone, vein quartz
Quartz-kyanite schist	Greenish white	Coarse-grained, crystalloblastic	Schistose	Quartz, kyanite	Muscovite	Argillaceous rock
Quartz-mica-schist	Greenish yellow with shades of pink	Coarse-grained, crystalloblastic	Schistose	Quartz, muscovite	—	Feldspathic sandstone
Schist	Variously coloured	Fine- to coarse-grained, lineation/foliation present	Schistose	Quartz, mica, chlorite, hornblende	Kyanite, feldspar etc.	Shale
Serpentinite	Greenish	Fine- to medium-grained	Schistose	Serpentine	Chlorite, talc, olivine, brucite	Basic and ultrabasic igneous rock
Slate	Light coloured, gray, soft	Fine-grained, foliated, slaty, cleaved	Schistose	Quartz, mica, sericite, chlorite	Feldspar	Shale, tuff
Soapstone (talc schist)	Grayish white, shining, soft	Medium- to coarse-grained	Schistose	Serpentine, talc, chlorite	Amphibole, quartz, magnesite, calcite	Ultrabasic igneous rock

Graphic Construction of ACF, AKF and AFM Diagrams

A variety of triangular diagrams are used to display the composition, variation and relationships in metamorphic rocks. The most widely used of them are ACF, AKF and AFM diagrams. The methods of construction of these diagrams are described below.

8.1 ACF DIAGRAM

The ACF diagram represents the composition of metamorphosed mafic rocks, shaly limestones and dolomites. The steps for ACF plot are given below.

   i. Divide the weight percentage of each oxide in the rock by its molecular weight to get the molecular proportion.

   ii. Set A = Molecular proportion of Al_2O_3 – Molecular proportion of K_2O – Molecular proportion of Na_2O [to account for K-feldspar and plagioclase]

   iii. Set C = Molecular proportion of CaO – 10/3 (Molecular proportion of P_2O_5) – Molecular proportion of CO_2 [to account for apatite and calcite]

   iv. Set F = Molecular proportion of FeO + Molecular proportion of MgO – Molecular proportion of TiO_2 – Molecular proportion of Fe_2O_3 [to account for ilmenite and magnetite]

   v. Determine the sum A + C + F and calculate the percentages of A, C and F.

   vi. A, C and F percentages are plotted in a triangular grid paper.

Example: *The chemical composition of a metamorphosed basalt is given in Table 8.1. Plot it in ACF diagram.*

TABLE 8.1: Chemical composition of a metamorphosed basalt					
Constituent	*Weight percent*	*Constituent*	*Weight percent*	*Constituent*	*Weight percent*
SiO_2	49.3	CaO	8.4	MnO	0.2
Al_2O_3	15.6	Na_2O	3.1	CO_2	0.1
Fe_2O_3	3.7	K_2O	1.5	SO_3	1.6
FeO	7.2	TiO_2	1.6	H_2O	0.7
MgO	6.7	P_2O_5	0.3	**Total**	100

Step 1. Molecular proportions of the constituent oxides are computed by dividing the weight percentage by corresponding molecular weight (Table 8.2).

Step 2. A = Molecular proportion of Al_2O_3.– Molecular proportion of K_2O – Molecular proportion of Na_2O = 0.153 – 0.016 – 0.050 = 0.087

Step 3. C = Molecular proportion of CaO – 10/3 (Molecular proportion of P_2O_5) – Molecular proportion of CO_2 = 0.150 – 10/3 (0.002) – 0.002 = 0.141

Step 4. F = Molecular proportion of FeO + Molecular proportion of MgO – Molecular proportion of TiO_2 – Molecular proportion of Fe_2O_3 = 0.100 + 0.168 – 0.02 – 0.023 = 0.225

Step 5. A + C + F = 0.087 + 0.141 + 0.225 = 0.453

Step 6. A = (0.087 ÷ 0.453) × 100 = 19
    C = (0.141 ÷ 0.453) × 100 = 31
    F = (0.225 ÷ 0.453) × 100 = 50

Step 7. The plotting is shown in Fig. 8.1, Line L_1, which is 19th line parallel with the base line (line opposite to A) represents A =19, Line L_2, which is 31st line parallel with AF side of the triangle opposite to C represents C = 31 and L_3, which is 50th line parallel with the side AC of the triangle opposite to F represents F = 50. The point of intersection of lines L_1, L_2 and L_3 represents the composition of the metamorphosed basalt.

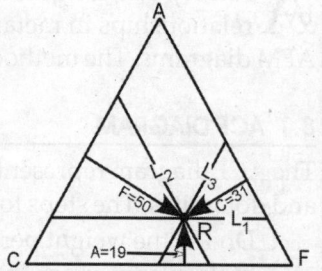

Fig. 8.1: ACF diagram

Constituent	Weight percent	Molecular weight	Molecular proportion
	TABLE 8.2: Calculation of molecular proportions of data given in Table 8.1		
SiO_2	49.3	60	0.822
Al_2O_3	15.6	102	0.153
Fe_2O_3	3.7	160	0.023
FeO	7.2	72	0.100
MgO	6.7	40	0.168
CaO	8.4	56	0.150
Na_2O	3.1	62	0.050
K_2O	1.5	94	0.016
TiO_2	1.6	80	0.020
P_2O_5	0.3	142	0.002
MnO	0.2	71	0.003
CO_2	0.1	44	0.002
SO_3	1.6	80	0.020
H_2O	0.7	18	0.039

8.2 AKF DIAGRAM

The AKF diagram is also used to represent the composition of metamorphic rocks. The steps for AKF plot are given below.

i. Divide the weight percentage of each oxide in the rock by its molecular weight to get the molecular proportion.

ii. Set A = Molecular proportion of Al_2O_3 + Molecular proportion of Fe_2O_3 – Molecular proportion of K_2O – Molecular proportion of Na_2O [to account for K-feldspar and plagioclase].

iii. Set K = molecular proportion of K_2O.

iv. Set F = Molecular proportion of FeO + Molecular proportion of MgO + Molecular proportion of MnO.

v. Determine the sum A + K + F and calculate the percentages of A, K and F.

vi. Plot A, K and F percentages in a triangular grid paper.

Example: *The chemical composition of a metamorphic rock is given in Table 8.3. Plot it in AFM diagram.*

Step 1. Molecular proportion of each constituent is computed by dividing the weight percentage by corresponding molecular weight (Table 8.4).

TABLE 8.3: Chemical composition of a metamorphic rock

Constituent	Weight percent	Constituent	Weight percent
SiO_2	42.8	TiO_2	1.6
Al_2O_3	18.9	P_2O_5	0.8
Fe_2O_3	3.9	MnO	0.6
FeO	4.8	CO_2	0.1
MgO	3.2	SO_3	0.6
CaO	9.3	C	0.8
Na_2O	9.6	H_2O	0.7
K_2O	2.3	Total	100

TABLE 8.4: Calculation of molecular proportions of data given in Table 8.3

Constituent	Weight percent	Molecular weight	Molecular proportion
SiO_2	42.8	60	0.713
Al_2O_3	18.9	102	0.185
Fe_2O_3	3.9	160	0.024
FeO	4.8	72	0.067
MgO	3.2	40	0.080
CaO	9.3	56	0.166
Na_2O	9.6	62	0.155
K_2O	2.3	94	0.024
TiO_2	1.6	80	0.020
P_2O_5	0.8	142	0.006
MnO	0.6	71	0.008
CO_2	0.1	44	0.002
SO_3	0.6	80	0.008
C	0.8	12	0.067
H_2O	0.7	18	0.039

Step 2. A = mol. prop. of Al_2O_3 + mol. prop. of Fe_2O_3 – mol. prop. of K_2O – mol. prop. of Na_2O
= 0.185 + 0.024 – 0.024 – 0.155 = 0.030

Step 3. K = mol. prop. of K_2O = 0.024

Step 4. F = mol. prop. of FeO + mol. prop. of MgO + mol. prop. of MnO = 0.067 + 0.080 + 0.008 = 0.155

Step 5. A + K + F = 0.030 + 0.024 + 0.155 = 0.209
$$A = (0.030 \div 0.209) \times 100 = 14$$
$$K = (0.024 \div 0.209) \times 100 = 12$$
$$F = (0.155 \div 0.209) \times 100 = 74$$

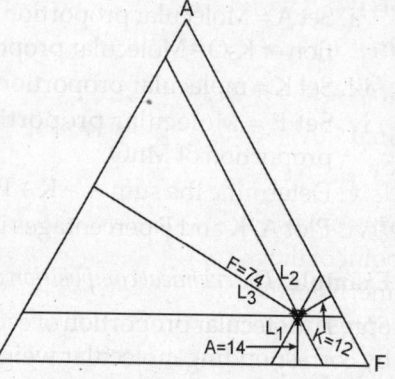

Step 6. The plotting is shown in Fig. 8.2. Line L_1, which is 14th line parallel with the base line KF (line opposite to A) represents A = 14, Line L_2, which is 12th line parallel with the side AF of the triangle opposite to K represents K= 12 and L_3, which is 74th line parallel with the side AK of the triangle opposite to F represents F = 74. The point of inter-section of lines L_1, L_2 and L_3 represent the composition of the metamorphic rock.

Fig. 8.2: AKF diagram

8.3 AFM DIAGRAM

Four major oxide components, viz. Al_2O_3, FeO, MgO and K_2O are responsible for much of the mineralogical variations present in pelitic rocks including index minerals like chlorite, biotite, garnet etc. These four components can be represented at the corners of a tetrahedron. A point inside the tetrahedron represents the bulk chemical composition of a pelitic rock. It is difficult to plot a three-dimensional tetrahedron on a two-dimensional sheet of paper. For simplification, the AFM face can be taken as the plane of projection as most of the characteristic minerals in pelitic rocks either lie on it or close to it. The steps for AFM plot are given below.

i. Divide the weight percentage of each oxide in the rock by its molecular weight to get the molecular proportion.

ii. Set A = molecular proportion of Al_2O_3 – 3 (Molecular proportion of K_2O)
[To project muscovite composition onto AFM face of AFMK tetrahedron]

iii. Set F = molecular proportion of FeO.
[If the rock contains ilmenite, set F = Molecular proportion of FeO – Molecular proportion of TiO_2]

iv. Set M = molecular proportion of MgO

v. Calculate the ratios A/(A + F + M) and M/(F + M)

vi. Plot the above ratios in a specially calibrated triangular grid paper as shown in Fig. 8.3. The M/(F + M) ratio is plotted on the base line, which is divided into 100 divisions. One small division of the original paper represents 0.01 of the ratio. Each divisional point is joined to the apical point of the triangular diagram. The A/(A + F + M) ratio is plotted on the left lateral side, which is divided into 16 major divisions—0.6 to 1.0. Thus, 62.5 small divisions (horizontal lines) of the original triangular diagram represent 1.0 of the ratio.

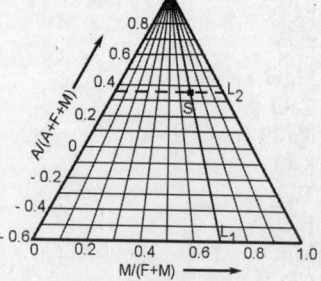

Fig. 8.3: AFM plot

Example: *The chemical composition of a shale that contains ilmenite is given in Table 8.5. Plot it in AFM diagram.*

Step 1. Molecular proportion of each constituent is computed by dividing the weight percentage by corresponding molecular weight (Table 8.6).

Step 2. A = mol. prop. of $Al_2O_3 - 3$ (mol. prop. of K_2O) = $0.151 - (3 \times 0.034) = 0.049$

Step 3. F = mol. prop. of FeO – mol. prop. of TiO_2 = $0.034 - 0.008 = 0.026$

Step 4. M = mol. prop. of MgO = 0.061

Step 5. A/(A + F + M) = $0.049 \div (0.049 + 0.026 + 0.061) = 0.049 \div 0.136 = 0.36$

M/(F + M) = $0.061 \div (0.026 + 0.061) = 0.061 \div 0.087 = 0.70$

Step 6. M/(F + M) = 0.70, so the line (L_1) joining the 70th division of the base line and the apex point of the triangle is marked. A/(A + F + M) = 0.36, so the $(0.6 + 0.36) \times 62.5 = 60$th horizontal line, from the base line towards apex (L_2) is marked. The intersection point (S) of these two lines represents the AFM plot of the given shale (Fig. 8.3).

TABLE 8.5: Chemical composition of a shale

Constituent	Weight percent	Constituent	Weight percent
SiO_2	58.10	TiO_2	0.65
Al_2O_3	15.40	P_2O_5	0.17
Fe_2O_3	4.02	MnO	0.00
FeO	2.45	CO_2	2.63
MgO	2.44	SO_3	0.64
CaO	3.11	C	0.80
Na_2O	1.30	H_2O	5.00
K_2O	3.24	Total	99.95

TABLE 8.6: Calculation of molecular proportions of data given in Table 8.5

Constituent	Weight percent	Molecular weight	Molecular proportion
SiO_2	58.10	60	0.968
Al_2O_3	15.40	102	0.151
Fe_2O_3	4.02	160	0.025
FeO	2.45	72	0.034
MgO	2.44	40	0.061
CaO	3.11	56	0.056
Na_2O	1.30	62	0.021
K_2O	3.24	94	0.034
TiO_2	0.65	80	0.008
P_2O_5	0.17	142	0.001
MnO	0.00	71	0.000
CO_2	2.63	44	0.060
SO_3	0.64	80	0.008
C	0.80	12	0.067
H_2O	5.00	18	0.278
Total	99.95		

8.4 USE OF ACF, AKF AND AFM DIAGRAMS

These diagrams are used to decipher mineral assemblages of metamorphic rocks derived from different parentages and formed in different physico-chemical conditions. The mineral assemblages characterize a particular metamorphic facies formed within a certain range of temperature and pressure condition. Let us assume that the rock considered for the construction of ACF diagram belongs to pyroxene-hornfels facies of contact metamorphism seen in the innermost zone of a contact aureoles. The principal mineral assemblages, which may contain both quartz and potash feldspar are as follows:

A. Pelitic and quartzo-feldspathic
  1. Quartz-orthoclase-andalusite-cordierite (-biotite)
  2. Quartz-orthoclase-plagioclase-andalusite-cordierite (-biotite)
  3. Quartz-orthoclase-plagioclase-cordierite (-biotite)
  4. Quartz-orthoclase-plagioclase-cordierite-hypersthene
B. Basic
  5. Plagioclase-hypersthene (-quartz)
  6. Plagioclase-diopside-hypersthene (-quartz)
  7. Plagioclase-diopside (-quartz)
C. Calcareous
  8. Plagioclase-diopside-grossularite
  9. Diopside-grossularite (-idocrase)
  10. Diopside-grossularite-wollastonite (-idocrase)

The rock (R) in the above mentioned ACF diagram example (Fig. 8.4) plots in the plagioclase-diopside-hypersthene (-quartz) field from which the mineral assemblages can be determined. AKF and AFM diagrams are used in similar manner.

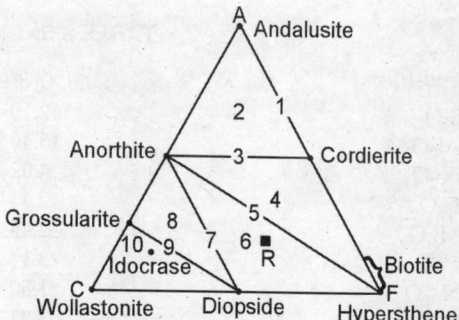

Fig. 8.4: ACF diagram of the metamorphosed basalt

Part III

Sedimentary Petrology

9. Texture of Sedimentary Rocks

10. Structure of Sedimentary Rocks

11. Classification of Sedimentary Rocks

12. Grain Size Analysis

13. Palaeocurrent and Palaeohydrological Analysis

14. Heavy Mineral Analysis

Texture of
Sedimentary Rocks

Rocks exposed to surface undergo weathering. As a result, the rocks disintegrate into smaller fragments and the constituent minerals are freed. A part of the rock decomposes forming soluble product. The weathered products, known as sediments, are transported to a site of deposition where they accumulate layer after layer. The soluble materials precipitate under favourable chemical conditions. A part of the sediments is contributed from volcanic source and another part, though insignificant in amount, is contributed from cosmic source. Subsequently, under low temperature and pressure condition prevailing at the near surface condition of the earth, diagenetic changes like compaction, cementation and lithification take place by which the sediments are converted to sedimentary rocks. Volumewise, these rocks constitute 5% of the lithosphere, but they cover more than 75% of the surface area of earth's crust.

The sedimentary rocks are characterised by their texture, structure and mineralogy. The texture refers to size, shape, roundness and arrangement of individual minerals of a rock. The structures, on the other hand, are large-scale features best seen in the field. These are produced due to dynamic effect of the transporting medium, chemical action and by the action of organisms. The mineralogical composition, texture and structure play vital roles in naming the rock. The constituents are clay and silt in case of shale and siltstone respectively; calcite, dolomite and fossils in case of limestone; quartz, feldspar, rock fragments and accessory heavy minerals in case of sandstones and larger rock fragments in case of conglomerate and breccia.

In case of clastic sedimentary rocks, texture includes size, shape and roundness of the constituent particles and their arrangement (fabric) and packing.

9.1 GRAIN SIZE

Unlike igneous rocks, constituent minerals (including rock fragments) of sedimentary rock are produced due to the effect of weathering. As a result, fragments more than meters in diameter to grains measurable in microns are produced. During transportation, abradation takes place that results in further breaking down of the grains. Thus, in contrast to igneous and metamorphic rocks, the grain size of sedimentary rocks varies within wider limits. To accommodate all the constituents it is pertinent to represent the grain size in logarithm scale proposed by Krumbain instead of a linear arithmetic scale. In this system the grain size is expressed as Phi '(ϕ)', which is the negative logarithm of the particle diameter in millimeters to the base 2.

$$\Phi = -\log_2 d \quad \text{and} \quad d = 2^{-\Phi}$$

Details of size analysis procedure and determination of size parameters are given in the "Grain size analysis" chapter.

9.2 SORTING

Sorting refers to the size range of the clastic particles constituting a sedimentary rock. The rock is said to be poorly sorted if the size range is large as in case of many conglomerates and breccias. Conversely, when the size range is narrow, the rock is said to be well sorted. The quantitative analysis of sorting has been dealt with in size analysis section. In case of very hard and indurated sandstones, in which the individual grains cannot be separated effectively, the approximate degree of sorting can be determined by microscopic observation. A visual comparison of the sandstone under microscope is made with Fig. 9.1 and the approximate qualitative and quantitative value of sorting can be determined.

9.3 SHAPE

The clastic sediments in the process of transportation are abraded with rounding of corners and tend to acquire an equidimensional shape. The limiting shape is that of a sphere, which is taken as reference in defining the shape of the grains. The shape (sphericity) of a grain is defined as the ratio of the surface area of a sphere (s) of the same volume as the grain under examination and the actual surface area of the grain (S).

$$\text{Sphericity } (\psi) = \frac{s}{S} \qquad \qquad ...(9.1)$$

The sphericity value of a sphere is 1.0 and for all other shapes the sphericity is less than 1.0. Due to the difficulty of measuring the surface area of an irregular solid, the sphericity is approximated by

$$\text{Sphericity } (\psi) = \frac{d_n}{D_s} \qquad \qquad ...(9.2)$$

where d_n = the diameter of the sphere of the same volume as the object
             D_s = the diameter of the circumscribing sphere, which is usually the greatest dimension of the sedimentary particle.

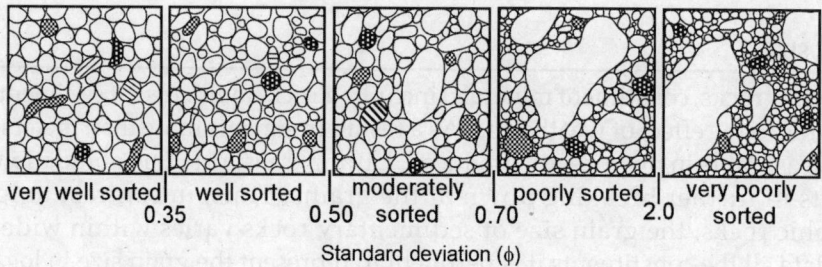

very well sorted | well sorted | moderately | poorly sorted | very poorly
0.35   0.50   sorted   0.70   2.0   sorted

Standard deviation (ϕ)

Fig. 9.1: Estimation of sorting of sandstone under microscope

The above mentioned equation is applicable for bigger size particles like cobble and coarse-grained pebble. The volume of the particle can be measured by immersing it in water taken in a graduated measuring cylinder from which the value of 'd_n' can be calculated and the greatest dimension 'D_s' can be measured by a slide caliper.

Example 1: *The volume of a pebble is 25 cm³ and the diameter of the circumscribing sphere is 5 cm. Determine the sphericity.*

Solution: Let the diameter of the sphere of the same volume as the pebble is d_n and the diameter of a circumscribing sphere is D_s.

Volume of equivalent sphere = $(\pi/6)\, d_n{}^3 = 25\ \text{cm}^3 \Rightarrow d_n = 3.6\ \text{cm}$

$$\text{Sphericity } (\psi) = \frac{d_n}{D_s} = \frac{3.6}{5} = 0.72$$

Determination of sphericity of particles of size smaller than pebble by the above mentioned method is difficult. In such cases, the projection sphericity is approximated to be equal to the true sphericity.

$$\text{Projection sphericity } (\psi_p) = \frac{dc}{DC} \qquad \qquad \text{...(9.3)}$$

where   dc = diameter of the circle equal in area to the grain projection area

DC = diameter of the smallest circle circumscribing the projected grain

In Fig. 9.2, the area of the grain is 2.177 unit $\Rightarrow$ dc = 1.665 unit; diameter of the circumscribing circle DC = 1.955 unit. So sphericity (Ψ_p) = 1.665 ÷ 1.955 = 0.85

Sahu and Patro (1970) referring to Riley (1940) define the projection sphericity as

$$\text{Projection sphericity } (\psi_p) = \frac{b}{a} \qquad \qquad \text{...(9.4)}$$

Where b = diameter of the largest inscribed circle and a = DC = diameter of the smallest circle circumscribing the projected grain.

In Fig. 9.3, b = diameter of the largest inscribed circle = 1.380 unit and length a = 1.995 unit

In this method sphericity $(\psi_p) = \dfrac{b}{a} = \dfrac{1.380}{1.995} = 0.69$

For determination of projection sphericity, photographs of the bigger grains can be taken by

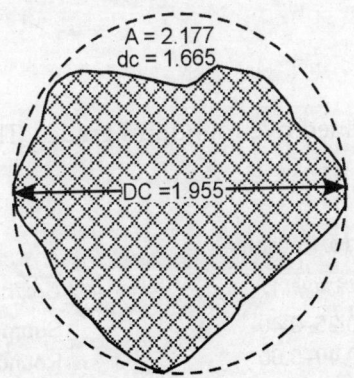

Fig. 9.2: Measurement of projection sphericity

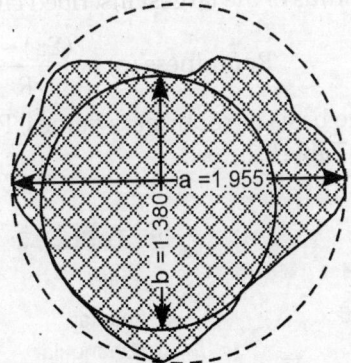

Fig. 9.3: Measurement of projection sphericity

TABLE 9.1: Zingg's shape classes

Class	b/a	c/b	Shape	Class	b/a	c/b	Shape
1.	$>2/3$	$<2/3$	Tabular or discoidal	3.	$<2/3$	$<2/3$	Triaxial or bladed
2.	$>2/3$	$>2/3$	Equant or spherical	4.	$<2/3$	$>2/3$	Prolate or rod-shaped

camera and photographs of the smaller grains can be taken by camera attached to the microscope eyepiece. The sphericity values of the grains in projection can be calculated by the formula (3) or (4). For more precision the average value can be taken into consideration.

Another way of defining shape of grains is based on the ratio of the length (a), breadth (b) and thickness (c) of the grain under consideration proposed by Zingg. This system is given in Table 9.1.

The sphericity index of sedimentary particles can be determined by the formula

$$\text{Sphericity } (\psi) = \sqrt[3]{\left(\frac{c^2}{ab}\right)}.$$

Example 2: *The length (a), breadth (b) and thickness (c) of a pebble are 6, 5 and 4 cm respectively. Calculate the sphericity index and determine Zingg's shape of the pebble.*

Solution: The sphericity index of the pebble $= \sqrt[3]{\left(\frac{c^2}{ab}\right)} = \sqrt[3]{\left(\frac{4^2}{5\times 6}\right)} = 0.81$

$$\frac{b}{a} = \frac{5}{6} = 0.83 > 2/3 \,(0.67) \text{ and } \frac{c}{b} = \frac{4}{5} = 0.8 > 2/3 \,(0.67)$$

The pebble is equant or spherical in shape according to Zingg's classification scheme.

9.4 ROUNDNESS

Roundness is the expression of sharpness of the edges and corners of the clastic grain. It is defined as the ratio of the average radius of curvature of several corners to the radius of curvature of the largest inscribed sphere. However, in reality these parameters are difficult to measure. The three dimensional grain is projected on two dimension paper and the roundness is defined as the average radius of curvature of the corners (r_i) of the grain in projection divided by the radius of the largest inscribed circle (R).

$$\text{Roundness } (\rho) = \frac{(\Sigma r_i) \div N}{R}$$

On the basis of roundness values, Pettijohn has defined five roundness grades. These are given in Table 9.2.

TABLE 9.2: Roundness grades of Pettijohn

ρ value	Grade	ρ value	Grade
0.00–0.15	Angular	0.25–0.40	Subrounded
0.15–0.25	Subangular	0.40–0.60	Rounded
		0.60–1.00	Well-rounded

Example 4: *Two images of pebbles (A and B) are shown in Figs 9.4 and 9.5. Calculate the roundness value in each case and assign the roundness grade.*

Solution: In pebble A (Fig. 9.4), there are 8 corners ($r_1 - r_8$) with radii 0.097, 0.04, 0.06, 0.04, 0.03 0.228, 0.06 and 0.13 unit respectively.

The average radius (Σr_i) ÷ N = (0.097 + 0.04 + 0.06 + 0.04 + 0.03 + 0.228 + 0.06 + 0.13) ÷ 8

= 0.685 ÷ 8 = 0.0856 unit. Radius of the largest inscribed circle (R) = 0.576 unit

Roundness (ρ) = 0.0856 ÷ 0.576 = 0.1486 ⇒ the pebble A is angular.

In pebble B (Fig. 9.5), there are 4 corners ($r_1 - r_4$) with radii 0.332, 0.25, 0.25 and 0.274 unit respectively.

The average radius = (0.332 + 0.25 + 0.25 + 0.274) ÷ 4 = 1.106 ÷ 4 = 0.2765 unit. Radius of the largest inscribed circle (R) = 0.575 unit

Roundness (ρ) = 0.2765 ÷ 0.575 = 0.48 ⇒ the pebble B is rounded.

The roundness values of smaller grains can be determined under microscope also. In these cases the magnified view of the thin section is projected onto a screen and grain boundaries are drawn. Roundness value of each grain is determined by measuring r_i and R. An approximate qualitative and quantitative visual estimation of the roundness can also be made by comparing the grain under consideration with that shown in Fig. 9.6.

Though roundness and sphericity are theoretically different, in nature they influence each other to certain extent. Different roundness grades of both low- and high-sphericity grains are shown in Fig. 9.7.

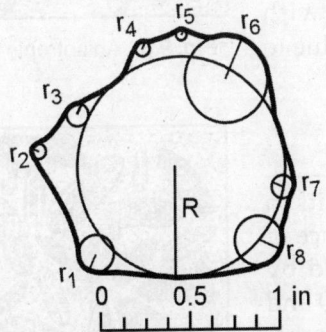

Fig. 9.4: Computation of roundness of pebble-A

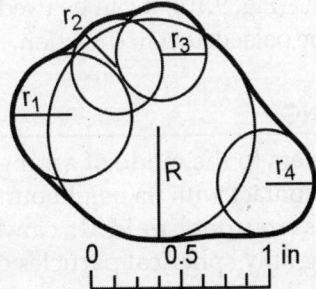

Fig. 9.5: Computation of roundness of pebble-B

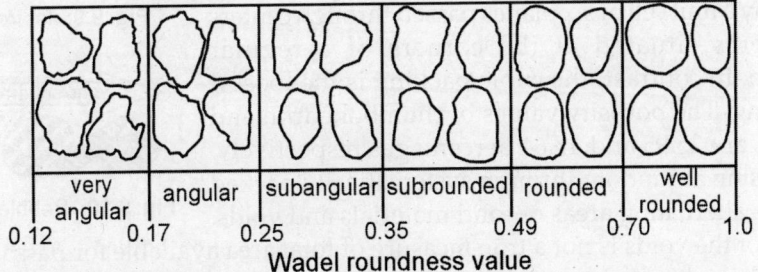

Fig. 9.6: Two-dimensional grain images for visual determination of roundness

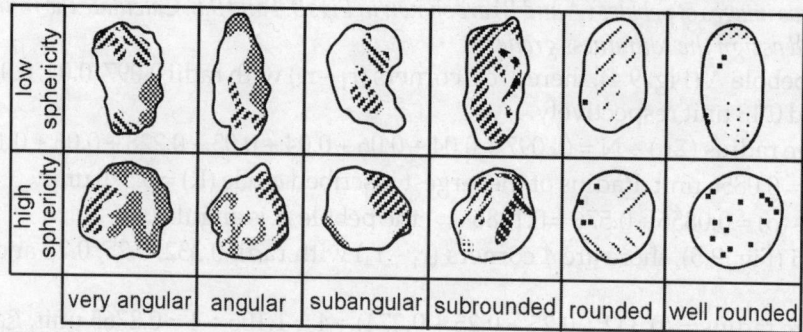

Fig. 9.7: Three-dimensional grain images for visual determination of roundness

9.5 FABRIC

The long axes of pebbles of conglomerates may show some degree of preferred alignment, which is termed anisotropic fabric (Fig. 9.8). If the long axes are randomly oriented, the fabric is termed isotropic (Fig. 9.9). In the former case, the preferred arrangement may be in response to hydrodynamic condition of the transporting medium. The preferred arrangement of pebbles in Fig. 9.8 may be due to current that was flowing either to north (top of figure) or to south (bottom of figure). However, if the alignment is associated with imbrications (Fig. 9.10), it can be used as an important clue to decipher the palaeocurrent direction.

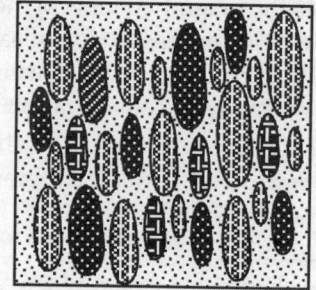

Fig. 9.8: Anisotropic fabric

9.6 PACKING

Packing refers to the mode of arrangement of solid units in tangential contact with its neighbours under the influence of earth's gravitational field. It can be well visualized by considering only spherical particles of sphericity of 0.80 and above. Ideally, six types of packings are possible, the cubic (Fig. 9.11) and rhombohedral (Fig. 9.12) being two extreme end members. The rhombohedral packing is the tightest being characterized by a unit cell of six planes passed through centers of eight spheres situated at the corners of a regular rhombohedron. In contrast, the cubic packing is the loosest possible packing. The porosity values of rhombohedron and cubic packings are 25.95 and 47.64 percentages respectively. Any plane passing at random through systematically packed spheres reveals alternating areas of solid materials and voids.

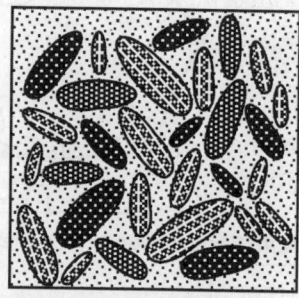

Fig. 9.9: Isotropic fabric

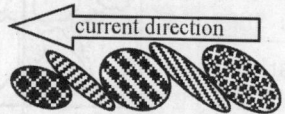

Fig. 9.10: Pebble imbrication

Thus, the area of the voids is not a true measure of total area available for passage of fluid as part of the voids is closed above or below by other grains which block the passage. A plane passing through the centers of the spheres in one of the rhobohedral layers (throat plane) is the

Fig. 9.11: Cubic packing

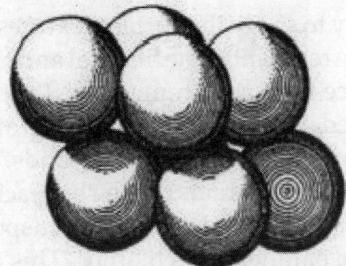

Fig. 9.12: Rhombohedral packing

true measure of the cross-sectional area for passage of fluid and thus known as 'useful or effective porosity'. In rhombohedral packing the total porosity is 25.95% and effective porosity is 9.30%. If large numbers of spheres of equal diameter are arranged in any manner, there is a certain diameter ratio for a small sphere which can just pass through the throats between large spheres into the interstices. For rhombohedral packing the critical diameter is 0.154D, where D is the diameter of large spheres. However, these theoretical concepts cannot be applied strictly in natural cases as the sedimentary particles are nonspherical and nonuniform in size.

Texture also describes the interrelationship between the grains that constitute the rock. Following textural terms are in vogue.

9.7 CLASTIC TEXTURE

The clastic sediments are primarily detrital in nature and lie in tangential contact with each other with intergranular pore spaces. When converted to rock, matrix and/or cement fill in the intergranular pore spaces partially or completely. The mosaic of grains and matrix with or without cement constitutes the clastic texture.

9.8 NONCLASTIC TEXTURE

Sediments precipitated by chemical or biochemical process from solution at the site of their accumulation show interlocking arrangement of grains without intergranular pore spaces. This type of texture is known as nonclastic texture.

9.9 ORGANIC TEXTURE

This type of texture is seen in rocks formed by accumulation of organic debris like shells, bones etc., which are well preserved and bounded mostly by finely divided organic matter.

9.10 SURFACE TEXTURE

Clastic sediments are carried by a medium, which may be ice, wind or water. During transportation, wear and tear of particles takes place by attrition and/or abrasion, as a result of which different types of marks are produced on the surface of the particles. These markings or imprints are known as surface texture. These are too small and insignificant in comparison to overall shape of the particles and difficult to be studied under petrological microscope with normal magnification. Scanning electron microscope (SEM) with high magnification is necessary to study the surface textures. Since quartz is the dominant mineral of the sediments and more resistant to mechanical and chemical wear and tear, it is commonly used for the study of surface textures. In many instances impact of one particle upon another gives rise to 'V' shaped depressions. The density of these depressions increases with wave turbulence and is maximum in case of deep-sea sands, which are deposited by turbidity currents. In contrast to marine deposits, grains broken by glacial ice exhibit characteristic conchoidal fracture surfaces. The surfaces of sand particles transported by wind are characterised by sharp and regular scratches parallel to each other. This type of mechanical abrasion of aeolian sand is termed *frosting*. Thus, the surface textures of quartz grains provide valuable clues for identification of their mode of transportation and environmental history.

9.11 MATRIX AND CEMENT

Matrix is the aggregate of smaller sized materials that fill in the inter-granular spaces in case of clastic sedimentary rocks and bind the grains. In matrix-supported conglomerates or breccias, sands fill in the inter-granular spaces forming arenaceous matrix. In many types of sandstone, matrix is constituted of fine silt, clay (argillaceous) or finely divided carbonaceous matters (carbonaceous), which are present in pore-fluid and/or deposited during diagenesis. The cement is chemically precipated material that binds the grains in clastic sedimentary rocks. It may be siliceous (SiO_2), ferruginous (Fe_2O_3) or calcareous ($CaCO_3$) in composition.

9.12 MICROSCOPIC STUDY OF DIAGENETIC FEATURES

The low temperature and pressure changes that affect the sediments when they are at or near the Earth's surface, so as to convert them to sedimentary rocks are known as diagenetic changes. Important diagenetic changes are lithification, compaction and cementation. Lithification includes those changes, which result in the formation of massive rock from loose sediment. Compaction involves the close packing of the individual grains by elimination of pore-spaces and expulsion of entrapped water caused by the weight of the overlying sediments. Cementation is the process by which the individual particles are held together by precipitated material, which may either be substances introduced by percolating ground water or derived from solution of a part of the mineral matter of the rock followed by redeposition.

The loose constituents of the conglomerates are bounded together by cement as it is done in case of many sandstones. In response to mechanical pressure, the pebbles are deformed with the development of microfaults (Fig. 9.13), which appear as minute step like displacements of the pebble surface. These features are well observed under microscope as well as in hand specimen.

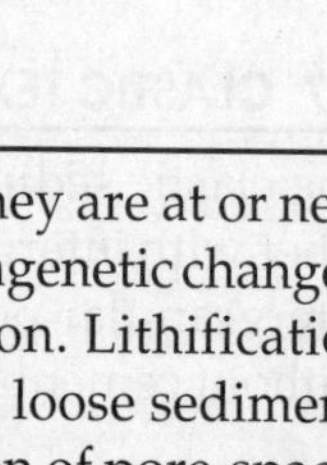

The effect of diagenesis, which may be either mechanical or chemical, is pronounced in case of sandstones. But in many instances they go side by side. Grain fracturing, bending and deformation of detrital mica and squashing of weaker grains are caused by mechanical effect. The chemical effects are solution, reprecipitation, decomposition and intergranular reactions. Redistribution of materials by solution at one place and precipitation at another place leads to cementation and reduction of pore space. The less stable framework grains are degraded and lose their identity and are

Fig. 9.13: Faulted pebble

transformed into microcrystalline matrix, which may interact with other more stable grains. The clay minerals in case of argillaceous matrix often undergo diagenetic change resulting in the formation of secondary mica. The net result of diagenesis is alteration of rock fabric, loss of porosity, blurring of original textures and transformation of the rock into most stable and equilibrium mineral assemblage. The floating grains come closer resulting in long-, point-, concavo-convex- and sutured-contacts (Fig. 9.14).

The most common cementing material is silica in the form of quartz, which is deposited as overgrowth on detrital quartz grains (Fig. 9.15). In many instances, optical continuity is maintained between the original and overgrown silica. In rare cases the silica is deposited in form of opal or chalcedony instead of quartz. In case of ferruginous sandstones, limonite derived from siderite constitutes the dominant cementing material. The effect of cementation, though to lesser extent, is also seen in case of overgrown feldspar (Fig. 9.16). In such case, cleavages crossing both the nucleus and secondary overgrowth are well marked, which point towards the optical continuity.

The porosity of freshly deposited carbonate rocks is comparatively very high of the order of 75 to 80 percent. Diagenetic effects like cementation and compaction reduce the original porosity by appreciable amount and result in the development of newly developed diagenetic fabric. It involves processes like cementation, cavity filling, grain growth and grain diminution.

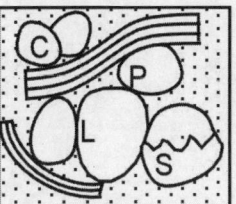

Fig. 9.14: Types of contacts: C. Concavo-convex, L. Long, P. Point, S. Sutured

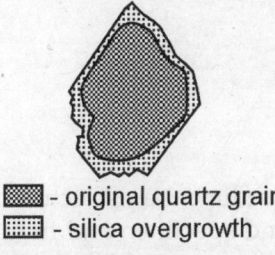

▓ - original quartz grain
▒ - silica overgrowth

Fig. 9.15: Overgrown quartz

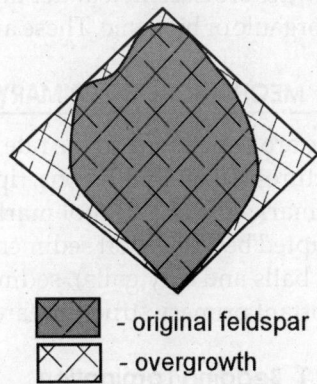

▨ - original feldspar
▧ - overgrowth

Fig. 9.16: Overgrown feldspar

10

Structure of Sedimentary Rocks

The sedimentary structures are large-scale features better observed in the outcrop than in hand specimen and in thin sections. Selectively collected hand specimens in proper orientation may exhibit the structures at least partially and can be studied in laboratory. The structures are classified under three groups, viz. mechanical or primary, chemical or secondary and organic or biogenic. These are described in following paragraphs.

10.1 MECHANICAL OR PRIMARY STRUCTURES

These are dependent on the rate of sedimentation and current of the medium. Planar bedding structures (beds, laminations, ripple marks, cross lamination, cross bedding, graded bedding), sole marks (flute and tool marks, groove casts, load casts and rain prints), deformed and disrupted bedding (soft sediment folding, boudinage, brecciations, mud crack casts, armored mud balls and clay galls), sedimentary sills and dykes belong to this category. Some of the important primary structures are described below.

10.1.1 Bedding/Lamination

The smallest sedimentation unit is the bed, which was deposited under essentially constant physical conditions. Since the current flow in nature is never absolutely uniform, no sediment is composed of particles of uniform size. The prevailing current of a particular mean velocity transports and deposits sediments of some particular size. When the current velocity changes drastically, a new set of conditions is established as a result of which sediments of differing grain size (often associated with changes in colour) are deposited. The beds are recognized and their top and bottom boundaries at a particular place are delineated on the basis of changes in size, colour and structure of sediments. A bed may be thick or thin or may be termed lamination depending on its thickness. The terminologies for stratified bed are given in Table 10.1. When the bed thickness is less than 1 cm it is termed lamination. Figure 10.1 shows 12 sedimentation units (laminations). The varves deposited by melt water of glaciers are good examples of laminations. Laminations are most characteristic of fine-grained sediments, particularly siltstones

Fig. 10.1: Laminations

TABLE 10.1: Terminology for stratification thickness

Thickness	Term	Thickness	Term
>1 m	Very thick bed	3–1 cm	Very thin bed
1 m – 30 cm	Thick bed	1 cm – 3 mm	Thick lamination
30–10 cm	Medium bed	<3 mm	Thin lamination
10–3 cm	Thin bed		

and shales. Light and dark colour of different lamina may be due to differences in organic matter or alternations of calcium carbonate and silt, etc.

10.1.2 Ripple Mark

Ripple marks are regularly spaced undulations on a sand surface or on a bedding plane of sandstone or coarse-grained siltstone. The wavelength is usually less than 50 cm and relief rarely exceeds 3 cm. Bed undulations exceeding these dimensions are referred to as dunes or sand waves. Ripples show wide variety of shapes each of which is related to a particular sedimentary process and hence are useful in interpretation of conditions of deposition. The ripples are of two types. The symmetric ripples (Fig. 10.2) are formed due to wave action in the bottom of standing water bodies. They may be either of peaked-crest type or rounded-crest type. The asymmetric ripples (Fig. 10.3) produced due to current action have a gentle stoss or up-current slope and a steep lee or down current slope and thus, are good indicators of palaeocurrent direction.

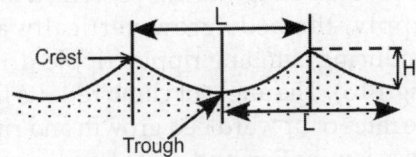

Fig. 10.2: Symmetric ripple

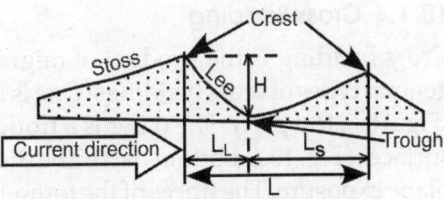

Fig. 10.3: Asymmetric ripple

The highest and lowest positions of the ripple surface are known as crest and trough respectively. The distance between two adjacent crests and troughs is the length (L) of the ripple and the difference between the elevations of crest and trough is the height (H) of the ripple. The length of the ripple can be resolved into two parts, viz. lee-side length (L_L) and stoss-side length (L_S). The ratio of length and height (L/H) is known as ripple index (RI) and the ratio of stoss-side length and lee-side length (L_S/L_L) is known as ripple symmetry index (RSI). These two parameters can be used to discriminate between ripples formed by wind, river and wave actions (Fig.10.4).

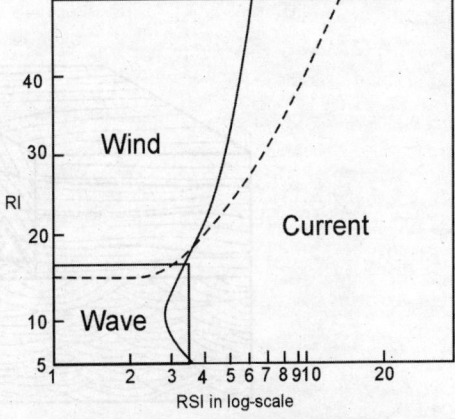

Fig. 10.4: Ripple discrimination diagram

10.1.3 Cross Lamination

Cross lamination is a type of internal lamination, which develops due to migration of ripples. It can be seen both on bedding planes and on vertical surfaces. Patterns of cross lamination are often specific to particular types of ripple and hence can also be used in interpretation of conditions of deposition. Ripples and cross lamination are principal features of sand grade sediments and also may be seen in coarse silts. They are most common in fine to medium grained sand and are rare in material coarser than coarse sand, except where they are due to wave action or strong winds.

Water movement over a sand bed as unidirectional currents, oscillatory waves or a combination of both may give rise to ripples and cross lamination. When the velocity of water flowing over a sand bed exceeds a certain critical value, grains begin to move. With a high rate of sediment supply, the beds grow vertically as ripples migrate producing climbing ripple (ripple-drift) cross lamination (Fig.10.5). The angle of climb reflects the balance between the rates of upward bed growth and ripple migration. The

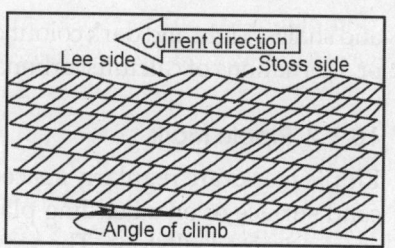

Fig. 10.5: Ripple-drift cross lamination

cross lamination and cross bedding are formed by migration of ripples and sand waves in response to prevailing current systems (but distinguished from each other by scale), they are used for determination of current direction and other flow parameters.

10.1.4 Cross Bedding

Cross bedding is the product of migration of a mega ripple or a sand wave. There are two general types of cross beddings. One is a simple tabular set with foresets approximately planes (Fig. 10.6, top part). The other is a trough shaped set of cross strata, which are usually curved surfaces (Fig. 10.6, bottom part). The distinction between these two types is made on a bedding plane exposure. The traces of the foresets in tabular cross bedding are straight lines, whereas in the second case they are remarkably curved and concave down current. The dip direction of foresets in case of tabular cross bedding and the bisertix of concavity in case of trough cross bedding indicate the direction of current flow.

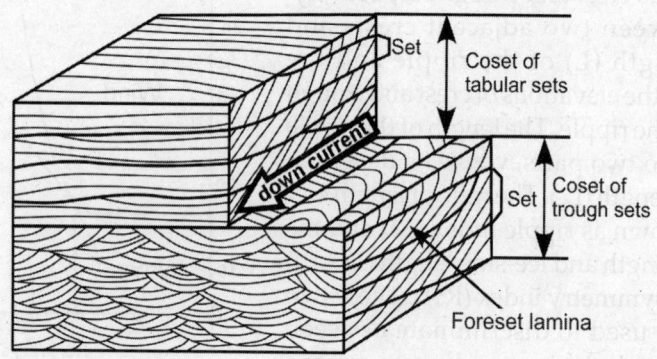

Fig. 10.6: Two principal types of cross bedding; tabular at top and trough at bottom

In case of cross beddings, the thickness of foreset layers is greater than 1 cm in thickness, whereas in case of cross laminations, the foreset layers are less than 1 cm in thickness.

Cross beddings commonly occur in sands of medium and coarser grain size. They may also occur in gravelly sands and fine gravels as well as conglomerates of any composition. Cross bedding is the product of migration of a sand wave whose size determines the scale of cross bedding. The regular linear ripples or sand waves produce the simple planar-tabular cross stratification while linguoid waveforms produce trough cross stratification. Cross bedding is one of the most widely used palaeocurrent indicators as large bed forms usually respond to a dominant flow and not easily remolded by low stage flow. The most valuable measurement in case of tabular cross bedding is the direction of dip of the foresets (foreset azimuth). The magnitude of dip is taken into consideration if the succession is tectonically tilted. To measure cross bedding in vertical sections, it is necessary to see faces in more than one orientation. The apparent dip on a single face only shows a component of true dip. A bedding surface view of the foresets gives the most accurate measurement of foreset azimuth. In case of trough cross bedding, it is necessary to measure the direction of trough axes on bedding planes. The scale of the cross bedding (foreset thickness, H in m) is a function ($H = 0.086\, d_s^{1.19}$) of water depth (d_s) above the sedimentary structure, i.e. mean water depth of stream in case of fluvial deposits. The deductions of channel parameters have been dealt with in palaeohydrology section.

10.1.5 Flaser, Wavy and Lenticular Beddings

In depositional system containing mud and sand, intermittent break in the current flow leads to preservation of mud streaks within ripple troughs. The resultant structure is known as flaser bedding (Fig. 10.7a). With increasing proportion of mud, wavy bedding (Fig. 10.7b) is produced. When the proportion of mud is very high in comparison to that of sand, i.e. lenticles of sand are preserved within thick layer of mud, the resulting structure is termed as lenticular bedding (Fig. 10.7c).

10.1.6 Swash Cross-stratification

Swash cross-stratifications are formed in the swash zone of beach. These are characterized by low-angle (2–10°) strata dipping towards the sea (Fig. 10.8). Individual sets of the swash cross strata are generally wedge shaped owing to variation in the beach slope that changes with unstable wave conditions. Laminae are subparallel to the lower set boundaries and generally show reverse textural grading.

Increasing proportion of mud

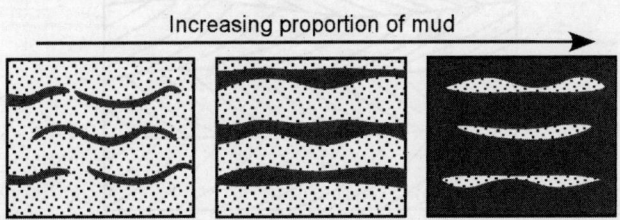

Fig. 10.7: (a) Flaser bedding, (b) Wavy bedding, (c) Lenticular bedding

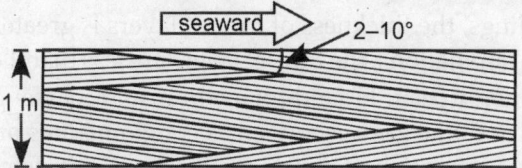

Fig. 10.8: Swash cross stratification

10.1.7 Hummocky Cross-stratification

Hummocky cross-stratifications consist of broadly undulating sets of gentle dipping (10–15°) laminae (Fig. 10.9) composed of fine sand and coarse silt. Dip directions of the individual sets show wide variation and cannot be used to deduce palaeoflow direction. Laminae parallel to lower set boundary are truncated by upper set boundary and thicken in dip direction. Hummocky cross stratifications are characteristic of marine shelf environment. These are possibly formed by sedimentation of suspended material over low hummocks and shallow swales produced by large storm waves.

10.1.8 Herringbone Cross-stratification

Herringbone cross-stratification is an example of bimodal and bipolar cross stratification in which two sets dip in exactly opposite directions observed in a single vertical section (Fig. 10.10). Formation of this type of cross stratification requires ebb and flood currents to have occurred at different times in a place where the rate of sedimentation is high enough to preserve the cross stratification. This type of stratification is the characteristic of tidal environment.

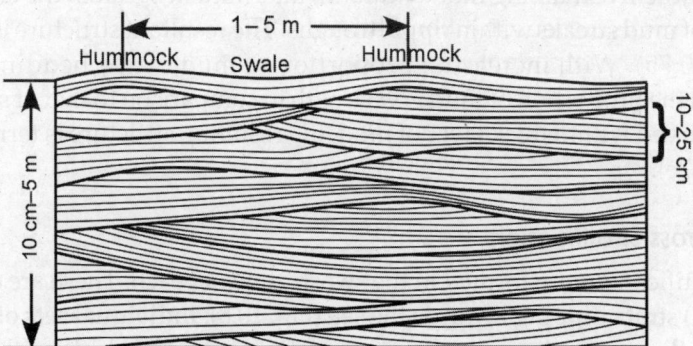

Fig. 10.9: Hummocky cross stratification

Fig. 10.10: Herringbone cross stratification

10.1.9 Graded Bedding

Graded beds are sedimentation units characterised by decrease in grain size from coarse to fine upward from the base to the top of the unit. Two types of graded beds exist. In one type, there is gradual gradation of particle size from the base to top, without any fines in the lowest part of the graded bed (Fig. 10.11). This type of graded bedding is possibly produced by waning current. In the second type there is overall decrease of grain size from base to top, but the fines are almost uniformly distributed (Fig. 10.12). These types of graded beds are characteristic features of turbidite deposition in water of considerable depth. The thickness of the graded unit ranges from less than a centimeter to more than a meter and the constituent materials are commonly sand and silt. In some cases, gravels may also take part in the formation of graded units. This primary sedimentary structure is useful in determination of top and bottom of beds and the order of superposition in isoclinally folded and overturned strata.

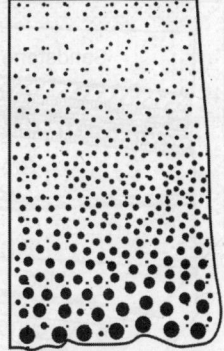

Fig. 10.11: Graded bedding produced by waning current

10.1.10 Sole Marks

Sole marks are bedding plane features present in the undersurface of some sandstone beds, which overlie shales. These are raised structures formed by the filling of depressions on the mud surface over which sand was deposited. These markings originate by current action and thus help in determination of palaeocurrent direction. Flute casts, groove casts and tool marks are some of the important sole marks.

  i. *Flute cast*: During flow of water a swarm of eddies develop and scour the mud surface. The size of scours depends on the coarseness of the materials carried by the stream and thus, is a function of the current strength. Infilling of these scours by loading of the sandstone deposited on the mud surface gives rise to protruding features, which are known as flute casts (Fig. 10.13). The flute casts

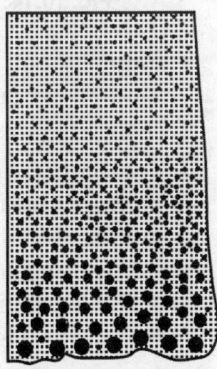

Fig. 10.12: Graded bedding produced by turbidity current

appear as raised structures on the base of the overlying sandstone. These are slightly elevated, elongate mound-like forms with bulbous upcurrent noses. The shape and size of the casts depend on the shape and size of the scours and commonly vary from a centimeter to more than a meter in length and a few millimeters to a few centimeters in height. The flute casts commonly occur in groups.

  ii. *Groove cast*: At times, furrows are engraved on the mud surface by dragging of objects like mud chips, shells, rock

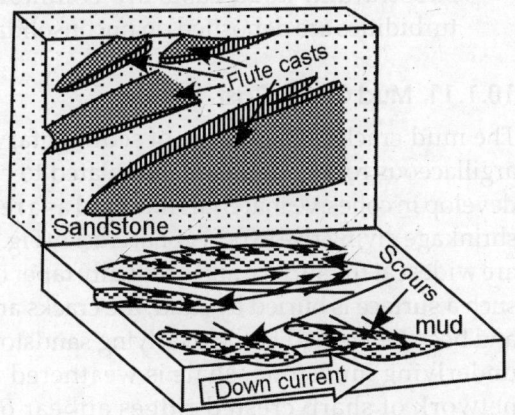

Fig. 10.13: Flute casts

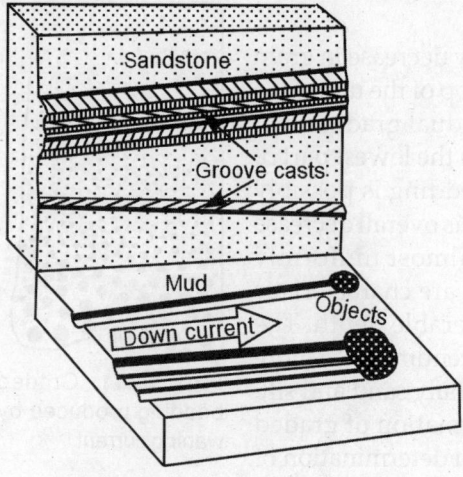

Fig. 10.14: Groove casts

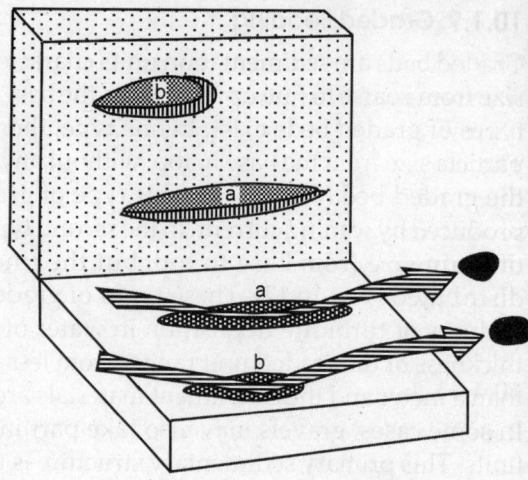

Fig. 10.15: Tool marks

fragments, etc. by the water current. Infilling of these furrows by sands deposited above mud gives rise to groove casts. The groove casts appear as rectilinear, rounded to sharp crested features on the base of the overlying sandstone bed (Fig. 10.14).

iii. *Tool mark*: Water currents carry various objects like mud chips, shells, rock fragments, etc. These objects move across the mud surface by rolling or intermittently impinging on the mud surface that creates scars. These scars are filled by sands deposited on the mud surface, which are preserved as small elevations on the base of the overlying sandstone. These features are known as tool marks (Fig. 10.15).

iv. *Load cast*: Load casts are irregular bulbous or mammillary features (Fig. 10.16) present in the base of sandstone beds underlain by shale or mudstones. These are caused by partial sinking of the heavy sand layer into a softer mud substratum. Load casts are commonly formed when turbidities are deposited on unconsolidated mud.

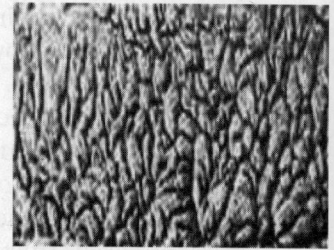

Fig. 10.16: Load cast

10.1.11 Mud Crack Cast

The mud crack cast is a type of sedimentary structure seen on the bedding planes of some argillaceous rocks. These are not related to current action and develop in cohesive materials like mud as a result of drying and shrinkage giving rise to polygonal cracks (Fig. 10.17). The cracks are widest at the surface and gradually taper downward. When such a surface is buried by sand, the cracks are filled with sand and become welded to the overlying sandstone bed. When the underlying mudstone/shale is weathered away, polygonal network of sharp crested ridges appear on the sole of the overlying sandstone.

Fig. 10.17: Mud crack casts

10.1.12 Rib and Furrow Structure

The rib and furrow structures (Fig. 10.18) are small and transverse markings, which occur in sets confined to long and narrow grooves separated from each other by narrow ridges. The narrow grooves are parallel to each other and to the direction of current flow. They are few centimeters wide and several meters in length. The small transverse markings within the grooves are arcuate, concave towards down current direction. Rib and furrow structures are generally associated with ripple marks.

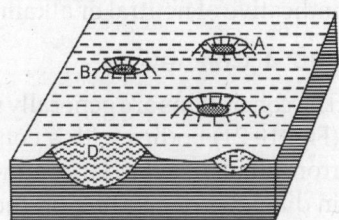

Fig. 10.18: Rib and furrow

10.1.13 Rain Prints

Rain prints are preserved in the same fashion as mud crack cast. Each print is a shallow depression encircled by a low ridge. In case of oblique rainfall, the elevation on the lee-side is higher than that on the windward side. Figure 10.19 shows three (A, B, C) raindrop prints in plan and two (D, E) in section. Impressions similar to rain print (hail, drip, spray, splash and bubbles) may also be found preserved as minor sedimentary structures.

Fig. 10.19: Raindrop prints

10.1.14 Armored Mud Balls

Armored mud balls are subspherical clay balls covered with fine gravels found embedded in sandstones. The size of the mud balls vary from 2–3 cm to about 50 cm in diameter, but those with 5–10 cm diameter are most common. The mud balls originate by release of clay chunks from riverbank, which upon rolling downstream acquire high sphericity and gravel cover. Though these are sparse in occurrence, provide valuable clues for estimation of palaeostream velocity and nature of bed material.

10.2 CHEMICAL OR SECONDARY STRUCTURES

These are produced by chemical action penecontemporaneous with sedimentation or shortly thereafter. Structures like nodule, spherulite, rosettes, concretions, veins, geodes, septaria, cone-in-cone, stylolite, corrosion zone, vug, oolicast, etc. belong to this category. Some of the important chemical structures are described below.

10.2.1 Nodule

Nodules are irregular tuberous (knobby) bodies (Fig. 10.20) commonly composed of chert and flint. These are devoid of internal structure and commonly aligned parallel to the bedding. These are produced due to post-depositional replacement of host rocks by silica.

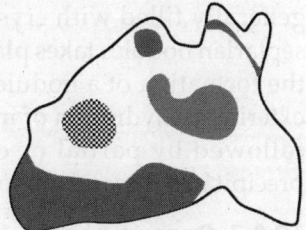

Fig. 10.20: Chert nodule

10.2.2 Spherulite

Spherulites are nearly spherical bodies, submicroscopic to a few centimeters in diameter. They are arranged radially around a center. The smaller spherulites resemble oolites and pisolites with radial symmetry. The spherical form is due to precipitation from colloidal gel state. The smaller spherulites are composed of chalcedony silica, apatite or aragonite, while larger varieties are carbonate concretions.

Fig. 10.21: Calcareous concretion

10.2.3 Rosette

These are accretionary bodies with symmetrical growth. They are made up of barite, marcasite or pyrite. Marcasite rosette is formed in acid (fresh water) environment while pyrite rosette is indicative of neutral or alkaline (marine) environment.

10.2.4 Concretions

The concretions are generally spherical, spheroidal or disc-shaped (Fig. 10.21) bodies with a concentric structure. They vary in size from small pellet to large spheroidal bodies of more than a meter in diameter depending on the permeability of the host rock. The concretions in sandstones are much larger in size than those found in shales. They are commonly composed of silica, calcite or iron oxide. Some of the concretions, particularly those composed of iron oxide are hollow (voidal) with a central void or cavity and outer dense limonitic layer (Fig. 10.22).

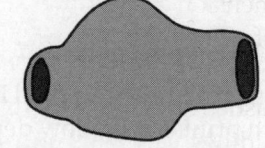

Fig. 10.22: Hollow iron oxide concretion

10.2.5 Geodes

Geodes are hollow globular bodies varying in size from a few centimeters to more than a meter in diameter. They are slightly flattened with their equatorial plane parallel to bedding. They are characterized by their subspherical shape, hollow interior, outer chalcedonic silica layer, inner drusy lining of inward pointing crystals. They are generally found in some limestone beds. A cross section of a geode is shown in Fig. 10.23.

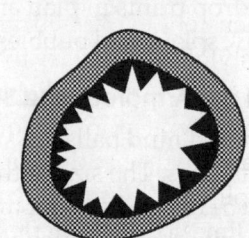

Fig. 10.23: Geode

10.2.6 Septaria

Septaria are large nodules characterized by a series of cracks that widen towards the center, which in turn are crossed by a network of cracks concentric with the margin (Fig. 10.24). The cracks are generally filled with crystalline calcite deposit. Formation of septarian nodules takes place in several stages. It is initiated with the formation of a nodule of aluminous gel, hardening of the exterior, dehydration of interior, formation of shrinkage cracks followed by partial or complete filling of the cracks with precipitated mineral matter which is commonly calcite.

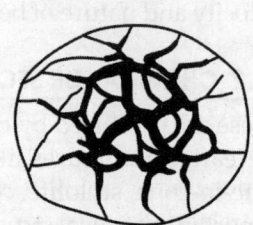

Fig. 10.24: Septaria

10.2.7 Cone-in-cone

Cone-in-cone is a minor structure noticed in some shales. These are commonly calcareous in composition, 1–15 cm thick and consist of a number of right circular cones (Fig. 10.25), which are

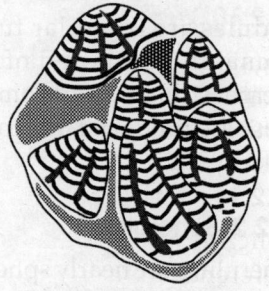

Fig. 10.25: Cone-in-cone

in inverted position with the cone axis normal to the bedding. The apical angles vary from 30° to 60° and the diameter of the circular base is about one-third of the height of the cone. The sides of the cone are generally ribbed or grooved and many are marked by depressions and ridges, which are pronounced near the base of the cone and become finer and obscure at the apex.

10.2.8 Stylolite

A stylolite seam is a surface discernible by interlocking or mutual interpenetration of two sides. The teeth like projections of one side fit into the sockets of exactly similar dimension on the other side. In cross section, the stylolite surface resembles the suture of cephalopods (Fig. 10.26). Sections of stylolite surfaces are well displayed on the polished surfaces of marble panels used in construction of buildings. The relief on a stylolite surface varies from less than a centimeter to more than two decimeters, the common amplitude being about a centimeter. The

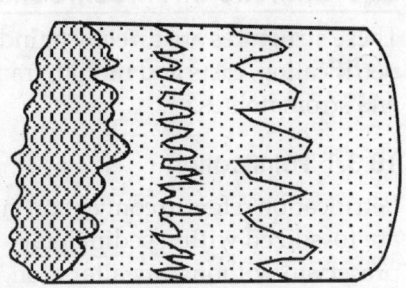

Fig. 10.26: Stylolite

dimensions of the teeth and socket correspond to the amplitude of the structure. The lengths of the stylolite seam vary from a few centimeters to several meters and commonly overlap or split into several subparallel branches. The stylolites of some sandstones contain parting of coaly matter, while those of quartzites are indicated by iron oxide. Commonly, the stylolites are parallel to the bedding but in a few cases, they are inclined or perpendicular to bedding planes. Stylolites occur in many types of rocks. However, they are most abundant in carbonate rocks like limestone, dolomite and marble. They have also been reported from sandstones, bedded siderites, gypsum, anhydrite, etc. The stylolites are formed by the action of pressure-solution in consolidated rocks. Since the formation of stylolites involves intrastratal solution, in few cases, it is possible to estimate the volume of rock dissolved in the production of stylolite seam.

10.2.9 Corrosion Surface

Corrosion surfaces are altered bedding surfaces of carbonate rocks formed by cessation of deposition of lime and removal of some of the previously deposited materials. Such a surface is characterized by minor irregularity and concentration of insoluble materials like quartz grains and phosphatic shells.

10.2.10 Vug

Vugs are irregular openings formed by the action of subsurface water. These are most common in carbonate rocks and also seen in some ferruginous sandstones. In many instances, they are partially filled with precipitated mineral matter.

10.2.11 Oolicast

Oolicasts are small subspherical openings produced by selective solution of oolites in preference to matrix in case of oolitic limestone. These structures enhance the porosity significantly facilitating the passage and movement of fluids.

10.2.12 Crystal mould

Crystal moulds are cavities in the rocks formed by the solution of salt, ice, etc. Salt testifies saline water and arid climate. The crystal moulds often imitate similar cavities formed by removal of pyrite cubes.

10.3 ORGANIC OR BIOGENIC STRUCTURES

These structures are directly or indirectly produced by organisms. Organic structures include petrification, casts and mould, tracks and trails, burrows and borings, fecal pellets, coprolites and stromatolites.

10.3.1 Petrification

Petrification refers to conversion into rock. The remains of organisms, under suitable condition are fossilized within sedimentary rock beds. The hard parts like bones, teeth, shells may be preserved without any alteration, but in many instances these remains are replaced by iron oxide, silica, carbonate, etc. The replacement takes place very slowly and in the process, both internal and external structures of the organism are found to be well preserved. In many instances, neither the original organic structure nor the replaced equivalent is present. In such cases, cavities are formed by removal of the original object in form of solution. These are known as moulds. The moulds show the outer form and ornamentation of the original object. When the mould is filled with foreign rock materials like silica, iron oxide etc, it is termed cast. A cast preserves the form and ornamentation but does not retain the internal structure of the replaced object. The plant remains, under favourable conditions, undergoes changes resulting in the enrichment of carbon. The process is known as carbonization and the end product is coal.

10.3.2 Tracks and Trails

Foot impressions left by birds and other animals are known as tracks. Typical tracks are elongate scratch marks (Fig. 10.27a), circular pits (Fig. 10.27b), bifid marks (Fig. 10.27c), imprints in groups (Fig. 10.27d) and clusters of small imprints (Fig. 10.27e). A track way is a succession of tracks. Common track ways are bifid and digitate marks (Fig. 10.27f) and oblique and longitudinal marks (Fig. 10.27g).

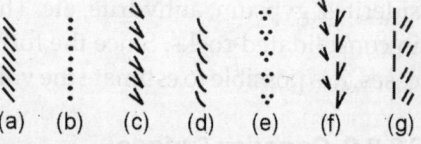

Fig. 10.27: Tracks and track ways: (a) Elongated scratch, (b) Circular pits, (c) Bifid mark, (d) Group imprint, (e) Cluster of small imprints, (f) Bifid and digitate marks, (g) Oblique and longitudinal marks

Continuous traces of grooves and furrows formed on the sediment surface by locomotion of organisms are known as trails. Different trail patterns are: gentle curving (Fig.10.28a), irregular meandering (Fig. 10.28b), zigzag meandering (Fig. 10.28c), meandering with highly attenuated loops in contact (Fig. 10.28d), spiral (Fig. 10.28e), and honeycomb network (Fig. 10.28f).

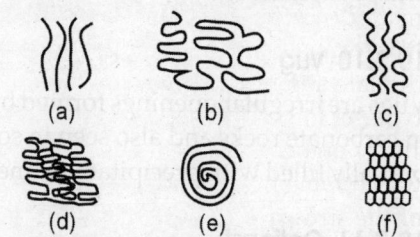

Fig. 10.28: Trail patterns: (a) Gentle curving, (b) Irregular meandering, (c) Zigzag meandering, (d) Meandering with loops, (e) Spiral, (f) Honeycomb

The trails may be of different shapes, e.g. simple with no ornamentation (Fig. 10.29a), simple with transverse ornamentation (Fig. 10.29b), bilobate with no ornamentation (Fig. 10.29c), bilobate with oblique ornamentation (Fig. 10.29d) and bilobate with transverse ornamentation (Fig. 10.29e).

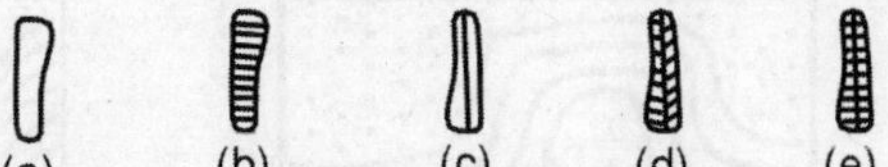

Fig. 10.29: Trail shapes: (a) Simple with no ornamentation, (b) Simple with transverse ornamentation, (c) Bilobate with no ornamentation, (d) Bilobate with oblique ornamentation, (e) Bilobate with transverse ornamentation

10.3.3 Burrows and Borings

Many organisms particularly mud eating worms make holes in soft sediments to live in. Burrows are excavations filled with sediments by active or passive filling after it is abandoned. Horizontal and nearly horizontal burrows are known as tunnels, whereas vertical or nearly vertical burrows are known as shafts. Borings are excavations in consolidated material such as lithified sediment, shell or bone.

10.3.4 Faecal Pellets

These are organic excreta of invertebrates present in modern marine deposits as well as in some sedimentary rocks. Nearly 30 to 50% of the sedimentary deposits are composed of these materials. They are rod shaped or ovoid with either longitudinal or transverse sculpturing or both in some cases. Simple ovoid forms with about one millimeter size are more common. The pellets are commonly transformed into glauconite, replaced by pyrite or may serve as centers for accumulation of phosphate.

10.3.5 Coprolites

These are large sized faecal pellets ranging in size 1–15 cm. These are characterized by dark colour, ovoid to elongate form and the surface may be marked by annular convolutions. Longitudinal striae and grooves are occasionally present. They are commonly phosphatic in composition.

10.3.6 Stromatolites

The stromatolite is a laminated structure composed of particulate sand-, silt- and clay-sized sediments, which has been formed by trapping and binding by algal mat. Since these are formed by blue-green algae, the term *algal stromatolite* is more appropriate. Generally the particulate matter is calcareous. The structure of stromatolites vary from flat lamination to small mound like forms of varying degree of complexity and size and simple columnar form to various digitate and branching forms. Stromatolites formed by different types of algae have different types of forms and crusts. The algal crusts may be flat and parallel to bedding (Figs 10.30 and 10.31), arched (Figs 10.32 and 10.33), shape of a stack of saucers (Fig. 10.34) and finger like forms, which split into two or more branches in upward direction (Fig. 10.35). The asymmetry of some of the stromatolites are useful in deduction of palaeocurrent and the upward convexity of laminations serve as a criterion to determine top of the strata and thus to determine the stratigraphic order in case of vertical or overturned beds.

The free-rolling forms of stromatolites are known as oncolites. Thrombolites are structures having a typical hemispherical external form but devoid of internal laminations.

Fig. 10.30: Weedia Walcott

Fig. 10.31: Archaeozoon Matthew

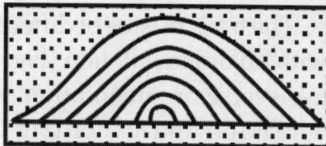

Fig. 10.32: Colenia Walcott

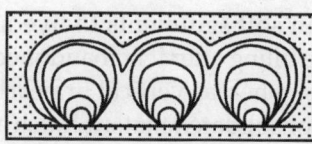

Fig. 10.33: Cryptozoon Hall

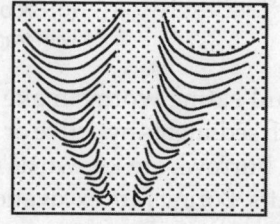

Fig. 10.34: Cryptozoon Boreale

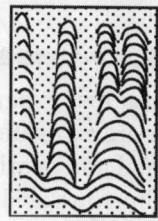

Fig. 10.35: Gymnosolen Steinmann

10.4 Trace Fossils or Ichnofossils

Trace fossils or ichnofossils are structures produced on the sediment surface by biogenic activity of organisms. In strict sense, these are produced by the behavioral activities of the organisms in response to ecological parameters and are not integral parts of organisms. Thus, burrows, borings, tracks, trails, etc. described above can be grouped under trace fossils. Trace fossils are abundant in sandstones and shales. Physical and/or chemical processes produce a number of sedimentary structures, which may resemble trace fossils. These structures are tool casts, flute casts, mud cracks, interference ripple marks, deformation structures, cone-in-cone, authigenic concretions, crystals and moulds of body fossils.

Different preservational terminologies are used to describe trace fossils. These are described below and shown in Figs 10.36 and 10.37.

i. *Full relief*: Traces preserved within a bed

ii. *Semirelief*: Traces at sand-shale contact

iii. *Epirelief*: Semireliefs on the top surface of sandstone bed. They may be concave up (negative epirelief) or convex up (positive epirelief).

iv. *Hyporelief*: Semireliefs on the bottom surface (sole) of sandstone bed. They may be concave down (negative hyporelief) or convex down (positive hyporelief).

v. *Spreite*: Spreite is a set of closely spaced, parallel or concentric traces. It can be either protrusive or retrusive (Fig. 10.37). Protrusive spreites are formed by the downward translocation of the vertical U-shaped burrow. The concave-down laminae are formed as sediment is

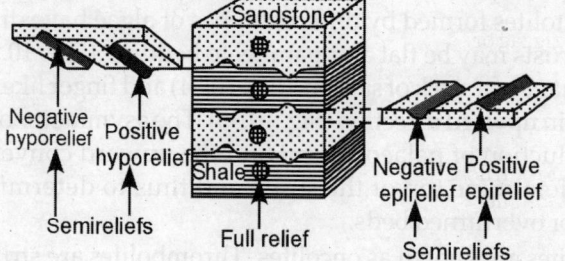

Fig. 10.36: Preservational terminology used to describe trace fossils

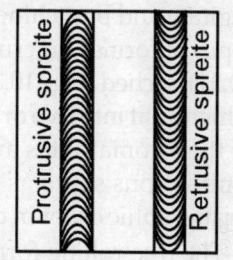

Fig. 10.37: Types of spreites

plastered onto the ceiling of the burrow. Retrusive spreites are produced by the upward trans-location of the vertical U-shaped burrow. Concave-up laminae are formed as sediment is plastered onto the floor of the burrow. The curved laminae are known as meniscae.

The full- and semi-reliefs are characterized by their shape, wall character, filling, spreite, size, orientation with respect to bedding, deformation of enclosing laminae, etc. Different types of these characters used to describe trace fossils are give in Table 10.2.

Descriptions of common trace fossils are given in Table 10.3 and illustrated in Figs 10.38 to 10.63.

TABLE 10.2: Morphologic characteristics used to describe trace fossils

I. Full-relief: Traces within stratum
A. Shape
  1. Unbranched
    i. Straight
    ii. Curved: (a) U-shaped, (b) J-shaped, (c) Spirally coiled
  2. Branched: (i) Regular, (ii) Irregular
  3. Wall diameter: (i) Uniform, (ii) Nonuniform
B. Wall character: (i) Unlined, (ii) Lined, (iii) Crenulated
C. Filling
  1. Homogenous
  2. Patterned: (i) Meniscate, (ii) Pelleted
D. Spreite: (i) Protrusive, (ii) Retrusive
E. Size
F. Orientation with respect to bedding: (i) Horizontal, (ii) Vertical, (iii) Inclined or oblique, (iv) Random
G. Deformation of enclosing laminae

II. Semi-relief
A. Epirelief or hyporelief: (i) Positive (convex), (ii) Negative (concave)
B. Shape
  1. Radially symmetrical
    i. With or without vertical structure
    ii. Circular or oval
    iii. Multilayered: (a) Shape of layers, (a) Number of layers
  2. Imprints on trackways
    i. Uniformity of imprints: (a) All alike, (a) Different kinds
    ii. Character of rows
      a. Continuous
      b. Discontinuous: Clusters/short rows oblique to trackways
    iii. Shape of imprints
      a. Simple: Elongated/circular or oval
      b. Digitate (number of digits)
  3. Ridges and furrows
    i. Pattern
      a. Gentle curving
      b. Meandering: No regular pattern/zigzag or sine curve/two orders of meanders/highly attenuated (meander loops in contact)
      c. Spiral
      d. Honeycomb network (regular hexagons)
    ii. Branched or unbranched
    iii. Shape: (a) Simple, (b) Lobate: Bilobate or trilobate
    iv. Ornamentation: (a) Transverse, (b) Oblique
C. Internal structure
D. Size
E. Orientation (relative to north)

TABLE 10.3: Descriptions of common trace fossils

Characters	Arenicolites (Fig. 10.38)	Asterosoma (Fig. 10.39)	Bergaueria (Fig. 10.40)	Bifungites (Fig. 10.41)
Position	Full relief	Positive hyporelief on sole of sandy beds	Full relief generally exposed	Positive hyporelief
Shape	Simple vertical U-shaped burrow without spreite	Elongate bulbous oval rays that branch radially from a central point; surface covered with longitudinal wrinkles	Stubby, clustered cylindrical traces with shallow depression on rounded blunt base often surrounded by 6–8 short radially arranged tubercles	Dumbbell or arrow-shaped
Internal structure	Generally smooth wall, but in some cases lined or sculptured	Concentric laminae of sand and clay packed central tube	Concentric sand-clay casing	None
Size	Distance between limbs of U is 1–10 cm; burrow diameter is 1–10 mm	Patterns contain three to nine rays and are 14–30 cm across; individual rays are 15–30 mm across and 30–80 mm long	Subequal length of 5–40 mm diameter	1–2 cm
Age	Cambrian to Holocene	Devonian to Cretaceous	Cambrian to Cretaceous	Cambrian to Devonian
Presumption	Dwelling burrow of suspension-feeding worms	Dwelling burrows made by decapod crustaceans	Dwelling trace of suspension-feeding coelenterate	Feeding burrow inhabited by small trilobites
Environment	Littoral to bathyal	Neritic and tidal flat	Neritic to bathyal	Littoral to bathyal

(Contd...)

TABLE 10.3: Descriptions of common trace fossils (Contd...)

Characters	Chondrites (Fig. 10.42)	Cruziana (Fig. 10.43)	Cylindrichnus (Fig. 10.44)	Diplichnites (Fig. 10.45)	Diplocraterion (Fig. 10.46)
Position	Full relief	Positive hyporelief	Full relief	Semirelief	Full relief
Shape	Three-dimensional branching straight cylindrical tunnels	Bilobate trails covered by V-shaped ridges having smooth or finely striated longitudinal zones outside V-marks	Elongated slightly curving subconical trace having variable orientation	Track comprising two parallel sets of fine ridges elongated at an angle with track axis	Vertical U-shaped burrow with spreite; limbs of U are parallel
Internal structure	None	None	Pinching out concentric sand-clay sheaths with central sand-filled tube	None	Longitudinal section shows several thin concentric laminae and horizontal section on bedding planes are dumbbell shaped
Size	0.5–5 mm diameter	Commonly 10–20 cm long and 0.5–8 cm wide, occasionally longer than 1 m	10–20 mm diameter with 2–4 mm central sand-filled tube	1–2 cm width track having 1–5 mm long fine ridges	Distance between limbs of U, 3–15 cm; depth of burrows, 2–60 cm; burrow diameter, 5–15 mm
Age	Ordovician to Holocene	Cambrian to Pennsylvanian	Mississippian to Cretaceous	Cambrian to Permian	Cambrian to Cretaceous
Presumption	Feeding burrow of sediment-eating worm	Trails made by ploughing movement of trilobites or similar arthropods	Dwelling burrow of filter feeder	Tracks produced by trilobites walking over muddy surface	Dwelling burrow of suspension feeder
Environment	Littoral to abyssal	Neritic	Neritic	Neritic	Littoral to neritic

(Contd...)

TABLE 10.3: Descriptions of common trace fossils (Contd...)

Characters	Helminthoida (Fig. 10.47)	Monocraterion (Fig. 10.48)	Nereites (Fig. 10.49)	Ophiomorpha (Fig. 10.50)
Position	Full relief	Full relief	Positive hyporelief	Full relief
Shape	Closely spaced parallel meandering tunnel trails	Vertical funnel-shaped	Closely spaced finely striated meandering structure consisting of narrow median furrow flanked by regularly spaced leaf-shaped lobes on both sides	Cylindrical, vertical and horizontal tunnels with local swelling and uneven outer surface
Internal	Horizontal paired tunnels structure with halo between, around, and below the tunnels	Downward warping of surrounding laminae toward the central tube; transverse section shows a series of concentric rings	None	Central gallery with meniscate laminae produced by active back filling
Size	1–3 mm wide tunnels. Length and width of meander are about 10 and 1 cm respectively	Tube diameter, 5 mm; length 8 cm; funnel diameter, 1–4 cm; funnel length, 2–4 cm	Width of trail, 1–2 cm	Tunnel diameter, 0.5–3 cm; length about 1 m
Age	Mississippian to Tertiary	Cambrian to Jurassic	Devonian to Cretaceous	Permian to Holocene
Presumption	Internal browsing trails of worms	Dwelling burrows of suspension-feeding worms	Internal grazing traces of gastropods, crustaceans etc.	Dwelling burrows of decapod crustaceans
Environment	Neritic to abyssal	Littoral	Abyssal to bathyal	Littoral (also upper neritic)

(Contd...)

TABLE 10.3: Descriptions of common trace fossils (*Contd...*)

Characters	Palaeophycus (Fig. 10.51)	Paleodictyon (Fig. 10.52)	Phycodes (Fig. 10.53)	Planolites (Fig. 10.54)
Position	Positive hyporelief	Positive hyporelief	Positive hyporelief	Full relief or positive hypo-relief
Shape	Cylindrical to subcylindrical sinuous burrows; horizontal or slightly oblique to bedding and commonly intersecting one another	Honeycomb like four- to eight-sided (generally hexagonal) polygonal network of ridges	Broom-like bundled horizontal tunnels	Horizontal, gently curved, unbranched cylindrical or subcylindrical infilled burrows commonly over-lapping each other
Internal structure	Sediment introduced by passive, gravity-induced sedimentation having the same composition as the surrounding matrix	None	May show spreite	Burrows filled with pelleted sediments of different composition, color and texture from the surrounding matrix
Size	Diameter, 3–15 mm; length about 20 cm	Width, 0.5–2 mm; mesh size ranges from 1 to 50 mm	Generally 15 cm long	Diameter ranges from 0.5 to 20 mm
Age	Precambrian to Holocene	Ordovician to Tertiary	Cambrian to Tertiary	Precambrian to Holocene
Presumption	Open burrow constructed by predaceous or suspension-feeding animal	Grazing traces formed at interface of sandy and muddy sediments	Feeding trace formed by sediment eating worms	Burrows of deposit-feeding animals
Environment	Neritic	Abyssal to bathyal	Neritic	Neritic

(Contd...)

TABLE 10.3: Descriptions of common trace fossils (*Contd...*)

Characters	Rhizocorallium (Fig. 10.55)	Rosselia (Fig. 10.56)	Scalarituba (Fig. 10.57)	Scolicia (Fig. 10.58)
Position	Full relief	Full relief	Full relief	Semirelief and full relief
Shape	U-shaped burrow oblique to bedding; upper part is vertical and lower part bends to sub-horizontal orientation; the outer surface of tube is often marked by numerous striations	Conical central sand-filled burrow that becomes horizontal at depth. The upper part may pass into other cones or flatten into Planolites	Subcylindrical sinuous burrows parallel, oblique, or normal to bedding	Band-like horizontal trilobate trail with varied sculpture
Internal structure	Longitudinal section shows long dark and light bands with parallel laminae or with spreite	Concentric sand-clay laminae that taper downward	The lower surface appears as a continuous series of bumps and the upper surface consists of a central furrow with a distinct meniscate structure with lateral lobes composed of oblique ridge furrows	None
Size	Burrow diameter varies from 1–3 cm; distance between the limbs of U ranges from 2–15 cm and length of U may be more than 70 cm	Diameter, 25–35 mm; height, 30–50 mm	Diameter, 2–10 mm; spacing between meniscae 2–3 mm	Width up to 4 cm; length up to 2 m
Age	Cambrian to Tertiary	Cambrian to Cretaceous	Ordovician to Permian	Cambrian to Tertiary
Presumption	Burrows of deposit-feeding animals like crustaceans or dwelling burrows of filter feeders	Dwelling or feeding burrow	Burrow made by worm or wormlike deposit feeder	Creeping or feeding trails of gastropods
Environment	Littoral to neritic	Neritic	Neritic to abyssal-bathyal	Neritic to abyssal-bathyal

(Contd...)

TABLE 10.3: Descriptions of common trace fossils (*Contd...*)

Characters	Scoyenia (Fig. 10.59)	Skolithos (Fig. 10.60)	Teichichnus (Fig. 10.61)
Position	Full or semirelief	Full relief	Full relief and positive hyporelief
Shape	Linear, curved, nonbranching burrow parallel, oblique or normal to bedding; covered by fine and clustered scratch marks	Straight, unbranched tubes perpendicular to bedding	Series of straight or slightly sinuous, non-branching and horizontal burrows stacked normal to bedding
Internal structure	Meniscate backfilled structure	None	The transverse section shows curved, concave or convex-up spreite and longitudinal section displays wavy long laminae (spreite) that merge upwards
Size	Diameter ranges from 1–10 mm	Diameter, 1–15 mm; length commonly up to 30 cm	Diameter, 3–20 mm; height up to 10 cm; length about 50 cm
Age	Permian to Holocene	Precambrian to Cretaceous	Cambrian to Holocene
Presumption	Burrows made by insect larvae, crustaceans or worms	Dwelling burrow inhabited by gregarious suspension-feeding wormlike animal	Burrows formed by deposit feeders
Environment	Littoral to neritic and also in nonmarine deposits especially red beds	Sandy littoral (also neritic to abyssal)	Neritic

(Contd...)

TABLE 10.3: Descriptions of common trace fossils (*Contd...*)

Characters	Thalassinoides (Fig.10.62)	Zoophycos (Fig.10.63)
Position	Full relief	Full relief
Shape	Cylindrical horizontal branch-ing network of burrows connected by vertical shafts swelling at branching points	Horizontal or inclined, circular, arcuate, or lobate spreite-filled loops at times forming a spiral around a vertical axis
Internal structure	Infilled as successive laminae which thin upward	Vertical sections show spreite structure
Size	Diameter ranges from 1 to 7 cm	Loop diameter about 60 cm and thickness ranges from 1–7 mm
Age	Permian to Holocene	Cambrian to Holocene
Presumption	Feeding and dwelling burrows or crustaceans	Feeding or grazing trace made by wormlike animal
Environment	Neritic	Neritic to abyssal

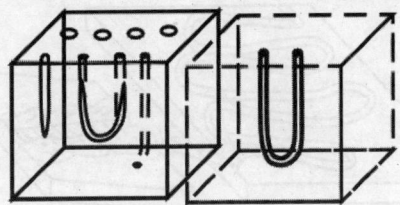

Fig. 10.38: Arenicolites

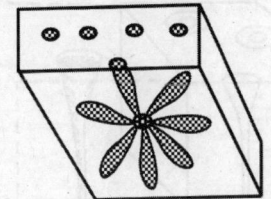

Fig. 10.39: Asterosoma

Fig. 10.40: Bergaueria

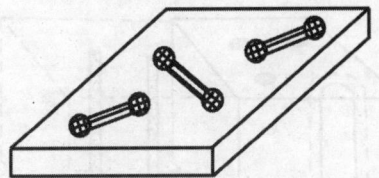

Fig. 10.41: Bifungites

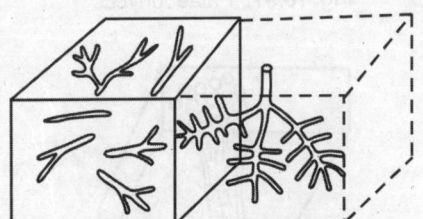

Fig. 10.42: Chondrites

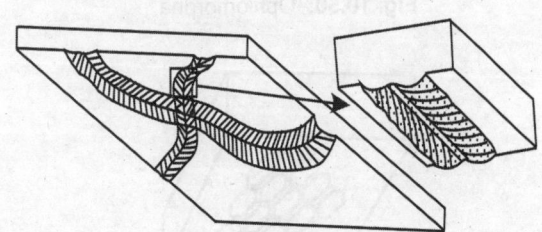

Fig. 10.43: Cruziana

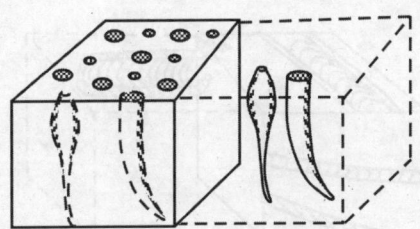

Fig. 10.44: Cylindrichnus

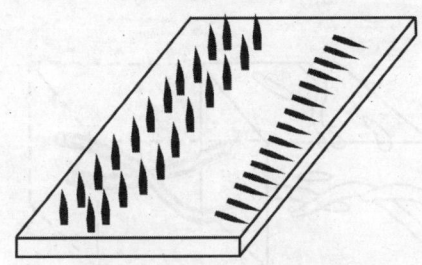

Fig. 10.45: Diplichnites

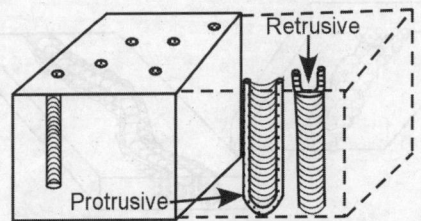

Fig. 10.46: Diplocraterion

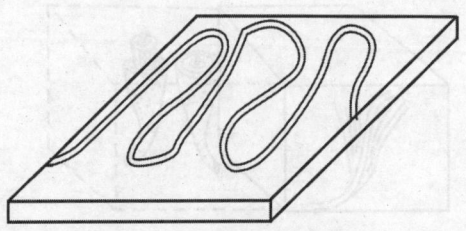

Fig. 10.47: Helminthoida

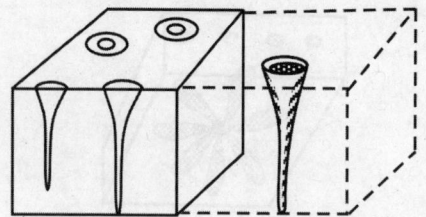

Fig. 10.48: Monocraterion

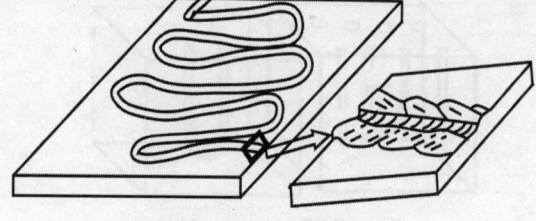

Fig. 10.49: Nereites

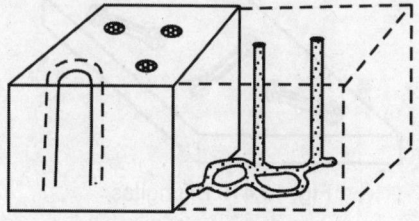

Fig. 10.50: Ophiomorpha

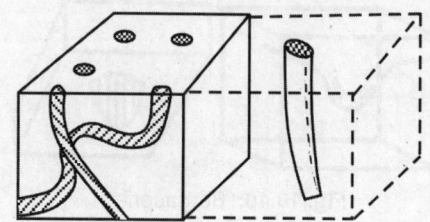

Fig. 10.51: Palaeophycus

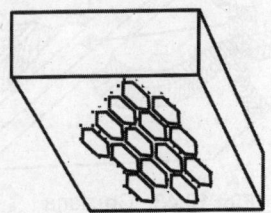

Fig. 10.52: Paleodictyon

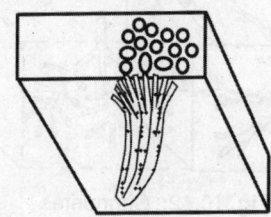

Fig. 10.53: Phycodes

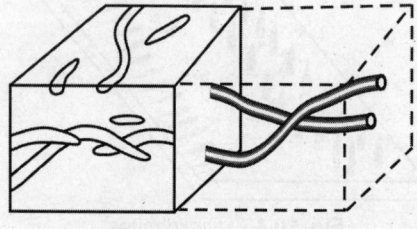

Fig. 10.54: Planolites

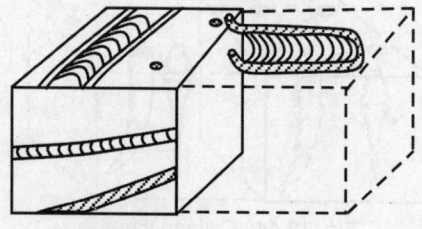

Fig. 10.55: Rhizocorallium

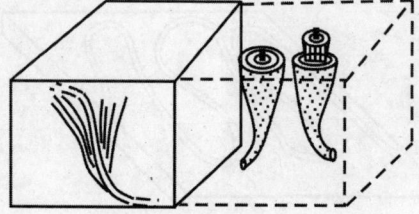

Fig. 10.56: Rosselia

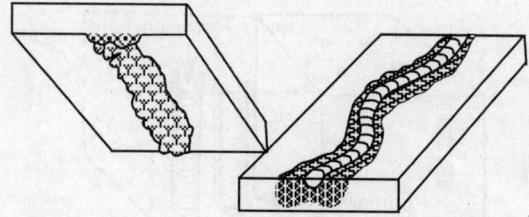

Fig. 10.57: Scalarituba

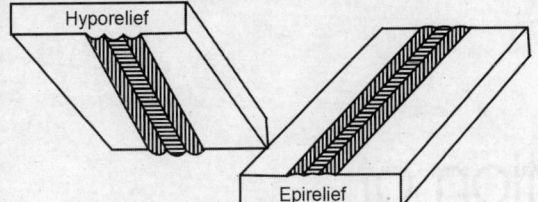

Fig. 10.58: Scolicia

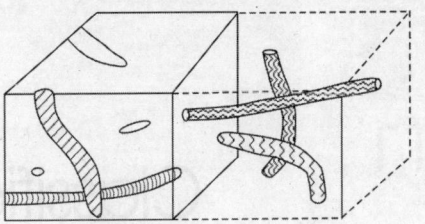

Fig. 10.59: Scoyenia

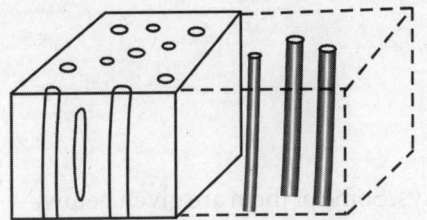

Fig. 10.60: Skolithos

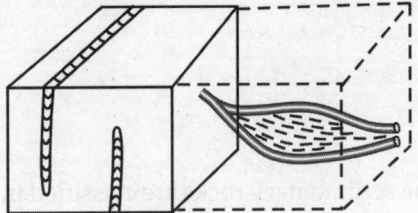

Fig. 10.61: Teichichnus

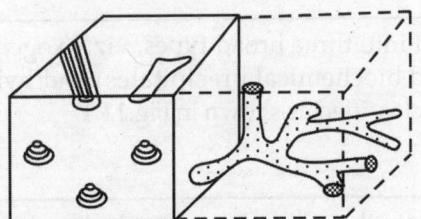

Fig. 10.62: Thalassinoides

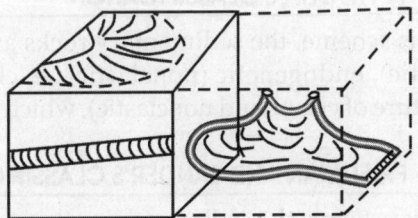

Fig. 10.63: Zoophycos

Classification of Sedimentary Rocks

The sedimentary rocks are classified in various ways. Some of them are given below.

11.1 PETTIJOHN'S CLASSIFICATION

In this scheme, the sedimentary rocks are classified into three broad types, viz. exogenetic (clastic), endogenetic (nonclastic, i.e. chemical and biochemical precipitates) and hybrid (mixture of clastic and nonclastic), which are further classified as shown in Fig.11.1.

11.2 FRIEDMAN AND SANDER'S CLASSIFICATION

In this scheme, the sedimentary rocks are classified into three broad groups, viz. intrabasinal (constituents are derived from within the basin of deposition), extrabasinal (constituents are derived from outside the basin of deposition) and pyroclastic (volcanic source). Their further classification is shown in Fig. 11.2.

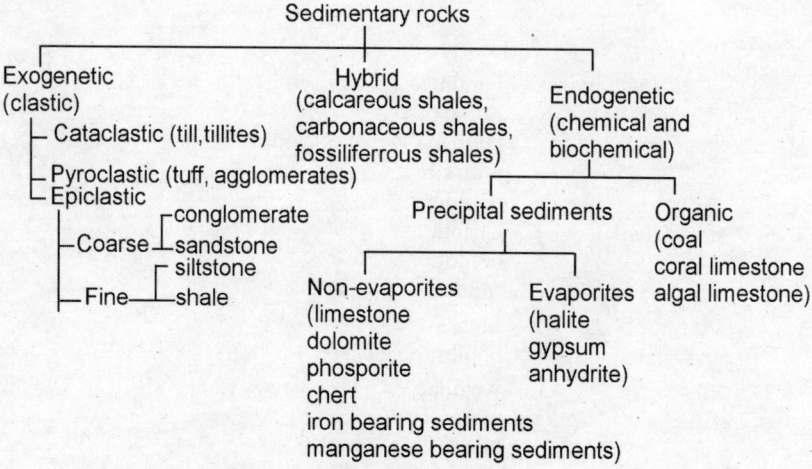

Fig. 11.1: Pettijohn's scheme of classification of sedimentary rocks

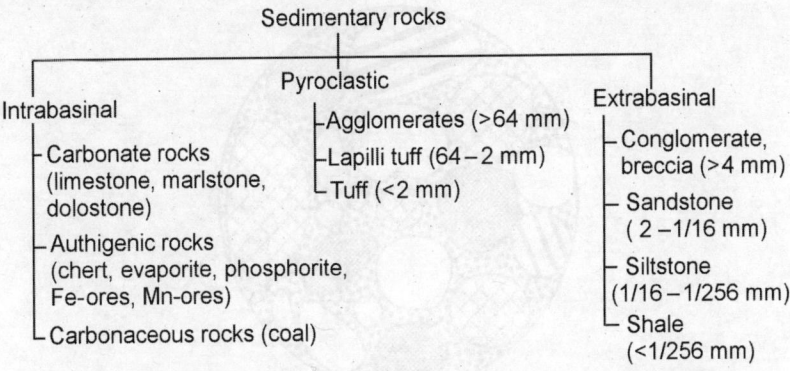

Fig. 11.2: Classification of sedimentary rocks by Friedman and Sander

11.3 CLASSIFICATION OF CONGLOMERATES AND BRECCIAS

Unconsolidated sediments coarser than 2 mm are grouped under the term gravel. Indurated gravels are called conglomerates. A breccia consists of gravel-size angular clasts. All breccias are not necessarily of sedimentary origin. Fault breccias are produced by tectonic activity. Large, well-rounded particles of volcanic origin (volcanic bombs) constitute a rock called agglomerate. The scheme of classification of conglomerates and breccias is given in Table 11.1. A thin section view of a conglomerate is shown in Fig. 11.3.

TABLE 11.1: Classification of conglomerates and breccias (modified after Sengupta, 2007)

Type of conglomerate	Name of conglomerate	Description	Origin and example
	Oligomictic	Pebbles of same composition, low matrix content	Derived from vein quartz, jasper or a monomineralic rock (generally fluvial or beach conglomerate)
	Polymictic	Pebbles of different composition, low matrix content	Derived from rocks of varying composition (generally fluvial or beach conglomerate)
Epiclastic	*Diamictite* — Conglomeratic argillite	Laminated argillites with ice rafted pebbles	Varves of glacial origin
	Diamictite — Tillite	High matrix conglomerates with striated, faceted, pentagonal boulders	Glacial
	Diamictite — Tilloid	Chaotic assemblage of non-glacial boulders in muddy matrix	Alluvial fan
	Intraformational	Mud or shale pebble intraclasts	Subaerial erosion
Pyroclastic	Volcanic breccia and agglomerate	Blocks of volcanic material and bombs solidified in flight	Volcanic
Cataclastic	Breccia	Angular fragments, crushed pieces, fault gouge, slickenside blocks	Slumping, folding, faulting, etc.
Meteoritic	Breccia	Shock effect noticeable	Meteoric impact

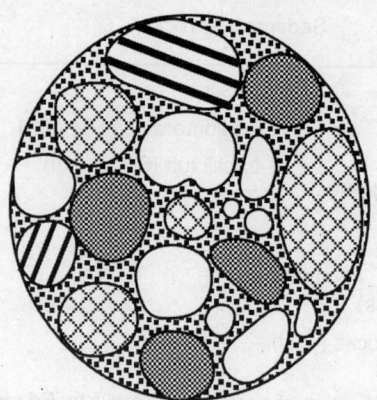

Fig. 11.3: Thin section view of a conglomerate

11.4 CLASSIFICATION OF SANDSTONES (DOTT, 1964; FOLK, 1968)

Quartz, feldspar and rock fragments are the basic constituents of sandstones. The sandstones are classified into different types depending on the relative proportion of these three end members. The detail scheme of classification in pictorial form is shown in Fig. 11.4.

Arrenite and wacke are two broad groups of sandstones differentiated on the basis of matrix and cement contents. Sandstones with less than 15% matrix and cement are termed arenite. These are relatively clean sandstones (Fig. 11.5). Wackes, on the other hand, contain more than 15% of matrix and cement. A thin section view of greywacke is shown in Fig. 11.6. In extreme cases, when the matrix and cement content exceed 75%, the sandstones grade into mudrocks.

11.5 CLASSIFICATION OF MUD ROCKS

Sediments of 0.063 to 0.004 mm size range are known as silt and sediments finer than silt are known as clay. Silt and clay together constitute mud. Indurated aggregate of silt and clay are known as siltstone and claystone respectively. Indurated, non-laminated mud forms mudstone. When the mudstone is laminated and fissile, the rock is designated as shale. Rocks composed of silt and clay and their mixture in various proportions are known as argillaceous rocks (mud rocks). The classification scheme is given in Table 11.2.

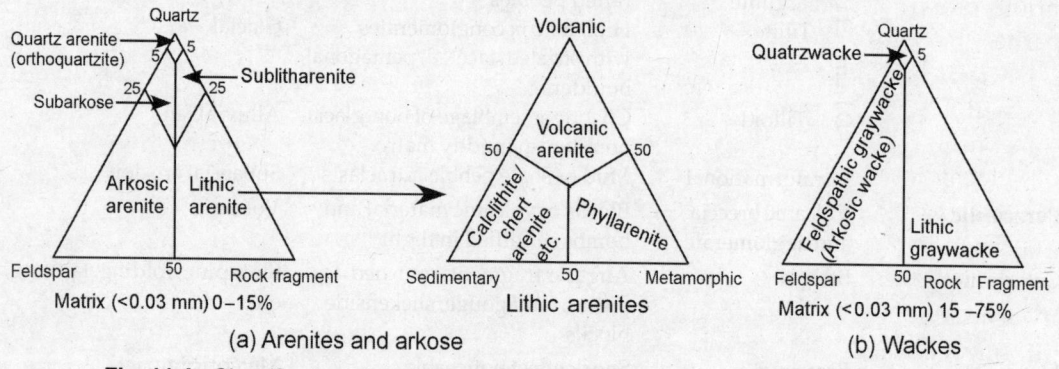

Fig. 11.4: Classification of sandstones by Dott (1964) and Folk (1968) in pictorial form

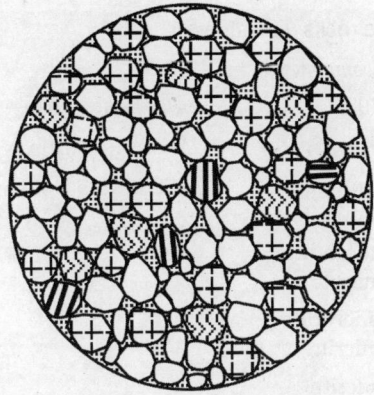

Fig. 11.5: Thin section of arenite

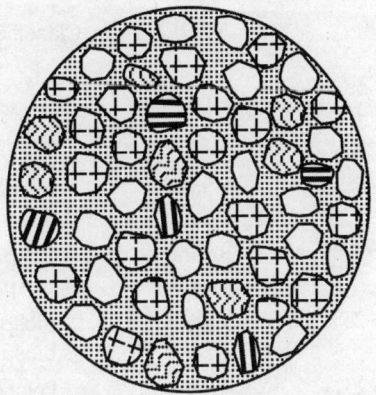

Fig. 11.6: Thin section of graywacke

TABLE 11.2: Classification of mudrocks

Percentage of clay-sized constituents	0–32	33–65	66–100
Feel to touch	Gritty	Loamy	Slick
Rock name	Siltstone	Mudstone	Clay stone

11.6 CLASSIFICATIONS OF CARBONATE ROCKS (LIMESTONE)

Limestones can be formed within the basin by precipitation from solution or by erosion of pre-existing carbonate deposits, transportation to the basin of deposition and accumulation like any other clastic sedimentary rocks. In the former case they are known as orthochemical (autochthonous) while in the later case they are known as allochemical (allochthonous). These two varieties are distinguished from each other by their texture and structure. The grains of the allochemical rocks are better sorted, their interstices are filled with calcite cement and they show structure like cross-beddings which support their transported nature. On the other hand, the grains of the orthochemical rocks are usually unsorted, the intergranular spaces are filled with lime mud and they show organic structure like undisturbed coral colonies, etc.

Most of the limestones are composed of orthochemical and allochemical components in various proportions. The orthochemical components are divided into two groups, micrite and sparite.

  i. *Micrite*: Micrite is microcrystalline calcite of very small size (0.001–0.004 mm) comparable with terrigeneous clay. In hand specimen, they are white, gray or black in colour and exhibit dull luster. They are formed by attrition of organic shells or deposited as biochemical ooze.

 ii. *Sparite*: Calcite crystals deposited within the pore-spaces are termed sparite. The crystals range in size from 0.01 to 0.1 mm, possibly formed by recrystallisation of micrite and generally occur as pore-filling material.

The allochemical constituents are intraclasts, fossils, oolites and pellets.

iii. *Intraclasts*: Intraclasts are fine-grained reworked fragments of pre-existing carbonate deposits.

TABLE 11.3: Classifications of carbonate rocks (limestones)

Dominant allochemical component	Grain size	Orthochemical component		Allochems < 10% Microcrystalline rocks
		Allochems > 10%		
		Sparite> Micrite	Micrite >Sparite	
Intraclasts > 25%	> Sand	Intrasparudite	Intramicrudite	Micrite and Dolomicrite
	Sand	Intrasparite	Intramicrite	
Oolites > 25%	> Sand	Oosparudite	Oomicrudite	
	Sand	Oosparite	Oomicrite	
Fossils > 25%	> Sand	Biosparudite	Biomicrudite	
	Sand	Biosparite	Biomicrite	
Pellets > 25%	< Sand	Pelsparite	Pelmicrite	

iv. **Fossils:** The fossils found in limestone are either entire or broken carbonate tests. Shells of brachiopods, molluscous, bryozoans, ecinoderms, corals, foraminifera, etc. are most common.

v. **Oolites:** Oolites are spherical carbonate bodies with concentric or radial structure. They vary in size from 0.1 to 1.0 mm.

vii. **Pellets:** Pellets are spherical to elliptical carbonate bodies without internal structure. They vary in size from 0.03 to 0.15 mm and are comparable with fecal pellets.

The limestones are classified on the basis of type and amount of allochemical and orthochemical components. The classification scheme proposed by Folk (1959) is given in Table 11.3.

11.7 CLASSIFICATION OF SEDIMENTARY AGGREGATES

Sedimentary aggregates are generally composed of materials of different size grades. Mixtures of sand and pebbles and sand-silt-clay (soil) are common in nature. There is no satisfactory agreement for the nomenclature of these mixtures. Classifications of pebble-sand by Pettijohn, gravel-sand-mud by Nicholas and sand-silt-clay by Shephard are given in Figs 11.7, 11.8 and 11.9 respectively. Nomenclature used for mixtures of terrigenous clastic sediments and sedimentary rocks are given in Table 11.4.

11.8 PETROGRAPHY OF SEDIMENTARY ROCKS

Petrographic descriptions of common sedimentary rocks are given in Table 11.5.

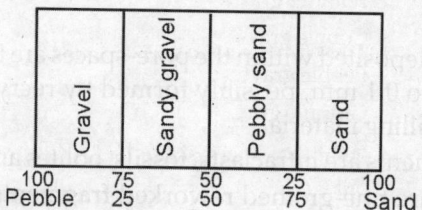

Fig. 11.7: Classification of pebble-sand mixture

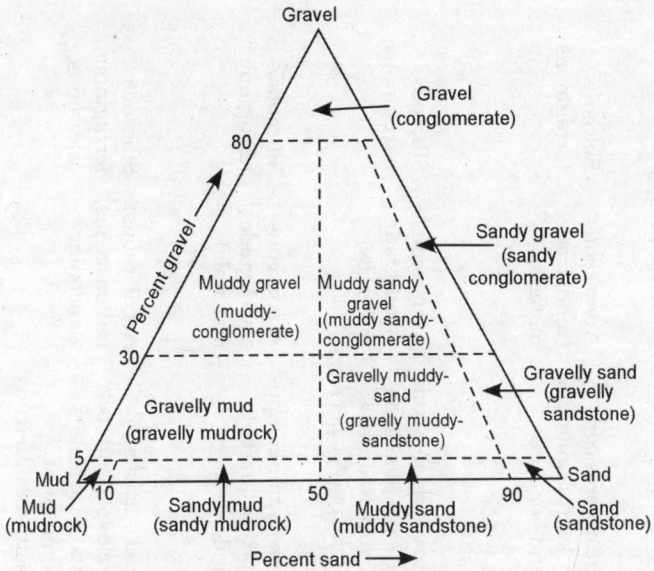

Fig. 11.8: Classification of gravel-sand-mud

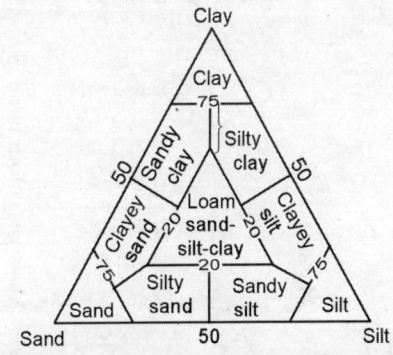

Fig. 11.9: Classification of sand-silt-clay

TABLE 11.4: Nomenclature used for mixtures of terrigenous clastic sediments and sedimentary rocks

Mean size		Rounded to subangular		Angular
(in mm)	Fragment /grain	Unconsolidated	Consolidated	Consolidated
>256	Boulder	Boulder gravel	Boulder conglomerate	—
256–64	Cobble	Cobble gravel	Cobble conglomerate	Rubble
64–4	Pebble	Pebble gravel	Pebble conglomerate	Breccia
4–2	Granule	Granule gravel		
2–1/16	Sand	Sand	Sandstone	1–½ mm, grit
1/16–1/256	Silt ⎱ Mud	Silt	Siltstone ⎱ Mudstone	
<1/256	Clay ⎰	Clay	Shale ⎰	

TABLE 11.5: Petrographic description of common sedimentary rocks

Rock	General character and texture	Structure	Essential minerals	Matrix	Cement
Polymictic conglomerate	Pebbles coarse-grained (>4 mm), rounded to subrounded, poorly sorted, isotropic/anisotropic, clastic texture, clast: matrix ratio is variable	Massive	Clasts are of variable composition–igneous/metamorphic rock fragments, quartz/feldspar/jasper/chert etc	Arenaceous (sand-grade materials)	Siliceous/ferruginous
Oligomictic conglomerate	Pebbles coarse-grained (>4 mm), rounded to subrounded, poorly sorted, isotropic/anisotropic, clastic texture, clast: matrix ratio is variable	Massive	Clasts are of same composition, such as quartzite, quartz and/or chert	Arenaceous (sand-grade materials)	Siliceous/ferruginous
Polymictic breccia	Pebbles coarse-grained (>4 mm), angular, poorly sorted, isotropic/anisotropic, clastic texture, clast: matrix ratio is variable	Massive	Clasts are of variable composition—igneous/metamorphic rock fragments, quartz/feldspar/jasper/chert etc.	Arenaceous (sand-grade materials)	Siliceous/ferruginous
Oligomictic breccia	Pebbles coarse-grained (>4 mm), angular, poorly sorted, isotropic/anisotropic, clastic texture, clast: matrix ratio is variable	Massive	Clasts are of same composition, such as quartzite, quartz and/or chert	Arenaceous (sand-grade materials)	Siliceous/ferruginous
Sandstone (exact type can be determined by modal analysis)	Variously coloured, coarse- to fine-grained (2–0.063 mm), angular to subrounded, poor to well sorted, clastic texture, clast: matrix ratio is high (>1)	Massive/colour banded/ripple mark/graded bedding/cross-bedding/cross-lamination/parallel lamination	Sand sized particles of quartz, feldspar and rock fragments of variable composition, mica, occasionally fossiliferrous	Argillaceous/carbonaceous/glauconitic	Siliceous/ferruginous/calcareous
Flagstone (micaceous sandstone)	Light coloured, coarse- to fine-grained (2–0.063 mm), angular to subrounded, poor to well sorted, clastic texture, clast: matrix ratio is high (>1)	Closely spaced bedding planes along which the rock splits, fissile	Quartz, muscovite, biotite, glauconite	Argillaceous/glauconitic	Siliceous/ferruginous/calcareous

(Contd...)

TABLE 11.5: Petrographic description of common sedimentary rocks (*Contd...*)

Rock	General character and texture	Structure	Essential minerals	Matrix	Cement
Shale	Variously coloured, fine-grained (<0.063 mm), clastic texture, clast: matrix ratio is high (>1)	Massive/ cross laminated, fissile, earthy smell when wet	Clay minerals, quartz, oxides of iron, manganese	Argillaceous	Absent/ ferruginous
Mudstone	Variously coloured, fine-grained, clastic texture, clast: matrix ratio is high (>1)	Massive/cross laminated, earthy smell when wet	Silt and clay sized grains of quartz, feldspar	Argillaceous	Absent
Limestone	Variously coloured, hard to soft, fine- to medium-grained, clastic/nonclastic texture, reacts with HCl	Massive/bedded/ripple marked/chemical structures/organic structures, pisolitic/oolitic	Calcite, dolomite, shell fragments or complete fossils	Micrite, sparite	Calcareous
Dolomite	Gray or white, fine- to medium-grained, granular, nonclastic texture, reacts with hot or conc. HCl	Massive/ bedded	Dolomite, calcite	Micrite, sparite	Calcareous
Marl	Gray, fine-grained, non-clastic texture, reacts with HCl	Massive / bedded, breaks with conchoidal fracture	Clay and carbonate minerals like calcite, dolomite	Micrite, sparite, argillaceous	Calcareous
Laterite	Various shades of brown	Massive, pisolitic, vesicular	Iron oxide, silica	—	Ferruginous
Marble	White/gray/dark coloured, nonclastic (granulose) reacts with dil HCl	Recrystallised	Calcite, dolomite, amphible, pyroxene, mica	—	
Coal	Dark gray to pitch black	Massive or banded	Vitrain (brilliant dark band), fusain (soft and powdery band), clarain (laminated bright band), durain (dull bands)	Carbonaceous	
Oosparite	Shades of gray, reacts with dil HCl, particles are rounded (fish row type)	Massive	Oolites (rounded, fish row type)	—	Spary calcite
Bio-sparudite	Gray, hard and compact, reacts with dil HCl, rudaceous	Massive	Allochems- fossils (foraminifera/brachiopod etc.)	—	Spary calcite
Bio-micrudite	Black, hard and compact, reacts with dil HCl, rudaceous	Massive	Allochems- fossils (foraminifera/brachiopod etc.)	—	Microcrystalline ooze
Micrite	Shades of pink, hard and compact, reacts with dil HCl, very fine grained	Massive	Microcrystalline ooze	—	Microcrystalline ooze

12

Grain Size Analysis

The dimensions of a large sedimentary particle can be conveniently measured by a slide calliper. If the particle is relatively smaller, the dimensions can be determined by micrometer scale fitted to the eyepiece of a petrological microscope. However, if the grain is very irregular, its volume can be found out by displacement of water and its radius (r) can be determined from the formula, volume $(V) = (4/3)\pi r^3$. The diameter determined in this process is known as *true nominal diameter*. Fragmental particles show wide variation of grain size ranging from boulder (> 25 cm) to finest clay (< 0.001 mm). On the basis of grain size, Udden and Wentworth classified the clastic sediments into different size grades. An ordinary arithmetic scale is not suitable to express the size ranges (grades). To overcome this difficulty, Krumbain introduced the method of logarithmic transformation, which is as follows:

$$\Phi = -\log_2 d \text{ and } d = 2^{-\Phi}$$

where, d = diameter of the particle in mm.

The scale so obtained is known as the 'phi (Φ) scale'.

In sedimentological laboratory, grain sizes of particles are determined by sieving and pipetting. In sieving a set of sieves of different mesh numbers are used. The mesh numbers indicate number of perforations per unit area. Thus, with increasing mesh number the size of perforation decreases. Three types of sieves, viz. BSS (British system), ASTM (American Standard of Testing Materials) and IS (International Standard) are available. The sieve mesh numbers and corresponding aperature sizes are given in Table 12.1. The ASTM sieve mesh numbers, mm- and Φ-sizes of sedimentary particles along with and Udden-Wentworth size class are given in Table 12.2.

12.1 METHOD OF GRAIN SIZE ANALYSIS

Grain size analysis of consolidated sedimentary rock includes disaggregation of grains, sieving, pipetting, conversion of weight values into percentage, drawing of simple and cumulative frequency curves, finding out of percentile values and computation of size parameters.

TABLE 12.1: BSS, ASTM and IS sieves with mesh number and aperture size
(IS sieves are indicated by their own aperture sizes)

Standard sieves with mesh number			Aperture size
BSS (410/1969)	ASTM (11-70)	IS (469/1972)	in mm
4	5	4.00 mm	4.00 mm
5	6	3.35 mm	3.36 mm
6	7	2.80 mm	2.83 mm
7	8	2.36 mm	2.38 mm
8	10	2.00 mm	2.00 mm
10	12	1.70 mm	1.68 mm
12	14	1.40 mm	1.41 mm
14	16	1.18 mm	1.19 mm
16	18	1.00 mm	1.00 mm
18	20	0.850 mm	0.84 mm
22	25	0.710 mm	0.71 mm
25	30	0.600 mm	0.59 mm
30	35	0.500 mm	0.50 mm
36	40	0.425 mm	0.42 mm
44	45	0.355 mm	0.35 mm
52	50	0.300 mm	0.30 mm
60	60	0.250 mm	0.25 mm
72	70	0.212 mm	0.21 mm
85	80	0.180 mm	0.177 mm
100	100	0.150 mm	0.149 mm
120	120	0.125 mm	0.125 mm
150	140	0.106 mm	0.105 mm
170	170	0.090 mm	0.088 mm
200	200	0.075 mm	0.074 mm
240	230	0.063 mm	0.063 mm
300	270	0.053 mm	0.053 mm
350	325	0.045 mm	0.044 mm
400	400	0.037 mm	0.037 mm
500	500	0.025 mm	0.031 mm

TABLE 12.2: Udden-Wentworth size classes of sedimentary particles

(ASTM) sieve mesh number	Phi (Φ)	Millimeter		Class	
	−12.00	4096.00		Boulder	
	−10.00	1024.00		(< −12 to −8 Φ)	
	−8.00	256.00			
	−7.00	128.00	Large	Cobble	
	−6.00	64.00	Small	(−8 to −6 Φ)	GRAVEL
	−4.00	16.00	Very coarse		
3	−3.00	8.00	Coarse	Pebble	
3.5	−2.50	5.66	Medium	(−6 to −2 Φ)	
5	−2.00	4.00	Fine		
6	−1.75	3.36			
7	−1.50	2.83		Granule	
8	−1.25	2.38		(2 to −1 Φ)	

TABLE 12.2: Udden-Wentworth size classes of sedimentary particles (*Contd.*)

(ASTM) sieve mesh number	Phi (Φ)	Millimeter		Class
10	−1.00	2.00		
12	−0.75	1.68	Very coarse	S
14	−0.50	1.41		
16	−0.25	1.19		
18	0.00	1.00		
20	0.25	0.84	Coarse	A
25	0.50	0.71		
30	0.75	0.59		
35	1.00	0.50		
40	1.25	0.42	Medium	N
45	1.50	0.35		
50	1.75	0.30		
60	2.00	0.25		
70	2.25	0.21	Fine	D
80	2.50	0.177		
100	2.75	0.149		
120	3.00	0.125		
140	3.25	0.105		(−1 to 4 Φ)
170	3.50	0.088	Very fine	
200	3.75	0.074		
230	4.00	0.063		
270	4.25	0.053	Coarse	S
325	4.50	0.044		
	4.75	0.037		
	5.00	0.031		
	5.25	0.026	Medium	I
	5.50	0.022		
	5.75	0.019		
	6.00	0.015		
	6.25	0.013	Fine	L
	6.50	0.011		
	6.75	0.009		
	7.00	0.008		T
	7.25	0.007	Very fine	
	7.50	0.006		
	7.75	0.005		(4 to 8 Φ)
	8.00	0.004		
	9.00	0.002	Coarse	
	10.00	0.001	Medium	CLAY
	<10.00	<0.001	Fine	(< 8 Φ)

12.1.1 Disaggregation of Grains

The unconsolidated sediments do not pose any problem for size analysis. In case of consolidated rocks, the grains are to be disaggregated first. Rocks with argillaceous and carbonaceous matrix do not create much problem. These rocks are broken into small chips followed by dis-aggregation by porcelain mortar and pestle. In case the non-calcareous grains bounded by calcareous cement, small chips of rock are treated with dilute HCl so that the cement is digested and grains are set free. Rocks with siliceous and ferruginous cement are heated up followed by sudden chilling. However, if the grains cannot be separated by any means, size can be determined by micrometer fitted microscope.

12.1.2 Sieving

Size analysis of pebble, granule and sand grade particles are conveniently determined by the sieving method. Thirty to seventy grams of dry sample is placed in the uppermost sieve in a set of stacked sieves with decreasing pore diameter, i.e. the coarsest sieve at the top with finer ones below. For example, ASTM sieves with mesh numbers 3, 3½, 5, 7, 10, 14, 18, 25, 35, 45, 60, 80, 120, 170, 230 and 325 are used if measurements at 0.5Φ intervals are required. As all the 16 sieves mentioned above cannot be accommodated in a single instance in sieve shaking machine, they are to be divided into two groups 3–25 mesh and 35–325 mesh. A lid is kept above the sieve of least mesh number to prevent jumping of grains during vibration of the sieves and a pan is kept at the bottom of the sieve stack to catch the grains which pass through the sieve of maximum mesh number. The shaker is run for 10–20 minutes. The materials retained in different sieves are carefully taken out and their weights are measured by digital balance. In many instances, silt and clay grade materials (< 325 mesh) are retained in the pan. The size analysis of this residue is determined by pipette analysis (sedimentation) method.

12.1.3 Pipette Analysis

The pipette analysis or sedimentation method is based on the settling velocity of silt and clay grade particles. The size of the particle, known as equivalent diameter (the diameter of a quartz sphere having the same settling velocity as the particle) is related to settling velocity by Stoke's law:

$$w = \frac{(\rho_s - \rho)g}{18\mu} \times d^2$$

where, w = settling velocity, ρ_s = density of particle, ρ = density of water, g = acceleration due to gravity (980 cm/sec^2), μ = dynamic viscosity of water (0.009) and d = equivalent diameter of the particle.

Materials required
1. 1000 ml graduated cylinder
2. 20 ml pipette
3. Stirrer
4. 50 ml beakers
5. Stopwatch
6. Low temperature oven
7. Sodium oxalate: 0.650 gm per 1000 ml water

Procedure

i. Take some water in a beaker and add 0.650 g of sodium oxalate and less than 325 mesh (4.5Φ = 0.044 mm) size sample to it. Stir with a glass rod till a homogeneous mixture is made. Sodium oxalate acts as an electrolyte to stop coagulation of fine particles.

ii. Transfer the mixture into a 1000 ml cylinder and add water till the mixture is made 1000 ml.

iii. Stir the mixture with the stirrer till it is made homogeneous. Stop stirring and start the stopwatch immediately.

iv. Suck out 20 ml of mixture by pipette at time intervals specified below from depths of 10 cm and keep in 50 ml beakers.

v. Heat the mixtures in low temperature oven till the water evaporates completely.

vi. Take out the residue, cool and weigh it by a digital balance.

Beaker	Time elapsed	Size of material	Beaker	Time elapsed	Size of material
1.	1 min 56 sec	<5.0Φ	5.	31 min	<7.0Φ
2.	3 min 52 sec	<5.5Φ	6.	1 hr 1 min	<7.5Φ
3.	7 min 44 sec	<6.0Φ	7.	2 hr 3 min	<8.0Φ
4.	15 min	<6.5Φ			

Computations

a. Say initial weight of the original material (<4.5Φ) is 15 g

b. Weight of the material including sodium oxalate in beaker – 1 is (< 5.0Φ) is 0.200 g

Weight of sodium oxalate in 20 ml mixture = 0.650 ÷ 50 = 0.013 g

Weight of material in 20 ml mixture is 0.200 – 0.013 = 0.187g

Total weight of <5.0Φ fraction in 1000 ml mixture is 0.187 × 50 = 9.35 g

Material in 4.5Φ – 5.0Φ range is 15.000 – 9.350 = 5.650 g

c. Weight of the material including sodium oxalate in beaker – 2 (< 5.5Φ) is 0.150 g

Weight of material in 20 ml mixture is 0.150 – 0.013 = 0.137 g

Total weight of < 5.5Φ fraction in 1000 ml mixture is 0.137 × 50 = 6.850 g

Material in 5 – 5.5Φ range is 9.350 – 6.850 = 2.500 g

So on…

The sandstones with siliceous and ferruginous cement are very hard and indurated. In these cases, the sandstones cannot be disaggregated and the grains cannot be separated effectively. In such cases, the sizes of the grains are determined under microscope. The microscope is fitted with measuring accessories by which grain diameter can be measured in different directions. In another method, a photomicrograph of the sandstone is taken and dimensions of each grain are measured in several directions taking the magnification factor into consideration. The average value of several measurements is the mean diameter of the concerned grain.

12.1.4 Analysis of Grain Size Data

The size analysis data is used to draw histogram and frequency curves. The histogram indicates the modal distribution of data while the percentile data obtained from cumulative frequency curve are used for computation of statistical parameters like mode, mean size, standard deviation, skewness and kurtosis by the following formulae proposed by Folk and Ward (1957).

$$\text{Median (Md)} = P_{50} = \Phi_{50}$$

where Φ_{50} is the 50th percentile value (P_{50}) obtained from cumulative frequency curve

$$\text{Graphic mean (Mz)} = \frac{P_{16} + P_{50} + P_{84}}{3} = \frac{\Phi_{16} + \Phi_{50} + \Phi_{84}}{3}$$

Mean size is used for classification of sediment into different grades (Table 12.2)

Inclusive graphic standard deviation (σ_1)

$$= \frac{P_{84} - P_{16}}{4} + \frac{P_{95} - P_5}{6.6} = \frac{\Phi_{84} - \Phi_{16}}{4} + \frac{\Phi_{95} - \Phi_5}{6.6}$$

Inclusive graphic skewness (SK_1)

$$= \frac{(P_{16} + P_{84} - 2P_{50})}{2(P_{84} - P_{16})} + \frac{(P_5 + P_{95} - 2P_{50})}{2(P_{95} - P_5)}$$

$$= \frac{(\Phi_{16} + \Phi_{84} - 2\Phi_{50})}{2(\Phi_{84} - \Phi_{16})} + \frac{(\Phi_5 + \Phi_{95} - 2\Phi_{50})}{2(\Phi_{95} - \Phi_5)}$$

$$\text{Graphic kurtosis (K}_G) = \frac{(P_{95} - P_5)}{2.44\,(P_{75} - P_{25})} = \frac{(\Phi_{95} - \Phi_5)}{2.44\,(\Phi_{75} - \Phi_{25})}$$

It is to be noted here that the Φ-values are the size values of corresponding percentiles in the cumulative weight percent graph. Φ_5, Φ_{16}, Φ_{25}, Φ_{50}, Φ_{75}, Φ_{84} and Φ_{95} are the size values corresponding to P_5, P_{16}, P_{25}, P_{50}, P_{75}, P_{84} and P_{95} percentile values respectively.

In addition to graphic measures indicated above, the grain size parameters can also be determined by moment measures by the formulae given below.

First moment (arithmetic mean): $\mu_1 = \dfrac{\sum (f_i m_i)}{\sum f_i}$

Second moment (variance): $\mu_2 = \sigma^2 = \dfrac{\sum (f_i m_i^2) - \sum (f_i m_i) \times \mu_1}{\sum f_i}$

Standard deviation (σ) $= \sqrt{\sigma^2}$

$$\text{Skewness} = \frac{\sum \left\{ (m_i - \mu_1)^3 \times f_i \right\}}{N \times \sigma^3}$$

$$\text{Kurtosis} = \frac{\sum \left\{ (m_i - \mu_1)^4 \times f_i \right\}}{N \times \sigma^4}$$

where m_i = midpoint of ith class; f_i = weight frequency of ith class; $N = \Sigma f_i$

The degree of sorting of the grains is inferred from the standard deviation values given in Table 12.3. Conclusion regarding relative dominance of the grain size of materials is determined from the skewness values (Table 12.4). The nature of size distribution is deduced from kurtosis values (Table 12.5).

TABLE 12.3: Inference of sorting of materials from standard deviation values

Standard deviation	Nature of sorting	Standard deviation	Nature of sorting
<0.35Φ	Very well sorted	0.71Φ to 1.00Φ	Moderately sorted
0.35Φ to 0.50Φ	Well sorted	1.00Φ to 2.00Φ	Poorly sorted
0.50Φ to 0.71Φ	Moderately well sorted	2.00Φ to 4.00Φ	Very poorly sorted
		>4.00Φ	Extremely poorly sorted

TABLE 12.4: Inference regarding relative dominance of the grain size of materials

Skewness value	Nature of sediment	Skewness value	Nature of sediment
+1.0 to +0.3	Very fine skewed	−0.1 to −0.3	Coarse skewed
+0.3 to +0.1	Fine skewed	−0.3 to −1.0	Very coarse skewed
+0.1 to −0.1	Near symmetrical		

TABLE 12.5: Inference regarding nature of size distribution from kurtosis values

Kurtosis value	Nature of distribution	Kurtosis value	Nature of distribution
< 0.67	Very platykurtic	1.11 to 1.50	Leptokurtic
0.67 to 0.90	Platykurtic	1.50 to 3.00	Very leptokurtic
0.90 to 1.11	Mesokurtic	> 3.00	Extremely leptokurtic

Example: *The grain size data of Karharbari sandstone (K-5) of the Ong-River Gondwana basin of Odisha obtained by sieving and pipette analysis is given in Table 12.6. The size parameters are to be determined.*

TABLE 12.6: Size analysis data of Karharbari sandstone K-5

Interval in Φ scale		Weight (Gram)	Weight percent	Cumulative weight percent
From	To			
	<−2	0.221	0.40	0.4
−2	−1.5	0.295	0.54	0.94
−1.5	−1.0	0.211	0.38	1.32
−1.0	−0.5	1.239	2.25	3.57
−0.5	−0.0	3.076	5.59	9.16
−0.0	0.5	6.849	12.45	21.62
0.5	1.0	7.533	13.69	35.31
1.0	1.5	10.858	19.74	55.05
1.5	2.0	8.963	16.29	71.34
2.0	2.5	3.643	6.62	77.97
2.5	3.0	3.506	6.37	84.34
3.0	3.5	1.222	2.22	86.56
3.5	4.0	2.261	4.11	90.67
4.0	4.5	2.45	4.45	95.12
4.5	5.0	0.231	0.42	95.54
5.0	5.5	0.55	1.00	96.54
5.5	6.0	0.55	1.00	97.54
6.0	6.5	0.25	0.45	98.00
6.5	7.0	0.2	0.36	98.36
7.0	7.5	0.35	0.64	99.00
7.5	8.0	0.3	0.55	99.55
	< 8.0	0.25	0.45	100.00

The histograms in arithmetic and logarithm ordinate scales are shown in Figs 12.1 and 12.2. It is seen that, in the later case, the visualization is better and polymodality of the size distribution data is clear. The cumulative weight percent data are also plotted in arithmetic- and probability-ordinate papers in Figs 12.3 and 12.4 respectively.

The cumulative frequency curve on arithmetic-ordinate graph paper is commonly S-shaped (Fig. 12.3). It provides the percentile values necessary for computation of size parameters. The cumulative frequency curve on probability-ordinate graph paper consists of segmented straight lines (Fig. 12.4). In addition to percentile values, it gives information about mode of transportation. The sediments of sandstone K-5 consists of traction population of size ranges -2Φ to -1Φ, three saltation populations of size ranges from -1Φ to 2Φ, 2Φ to 5Φ and 5Φ to 7Φ and suspension population of size range $>7\Phi$.

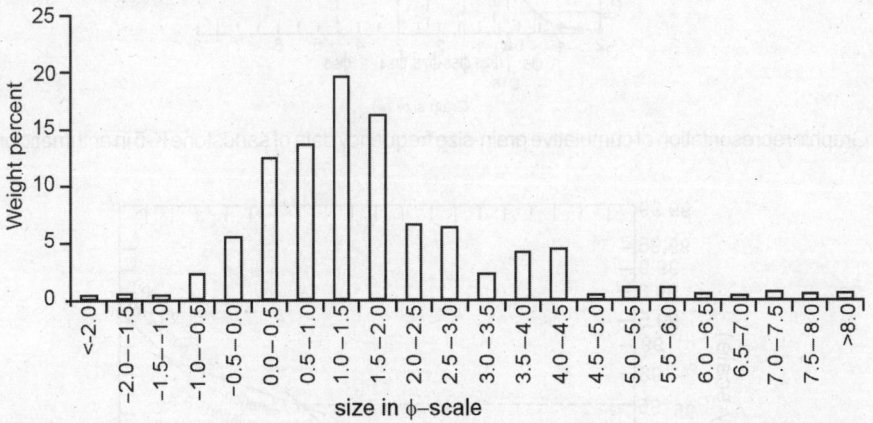

Fig. 12.1: Histogram of the grain size data of sandstone K-5 of the Ong-River Gondwana basin in arithmetic ordinate scale

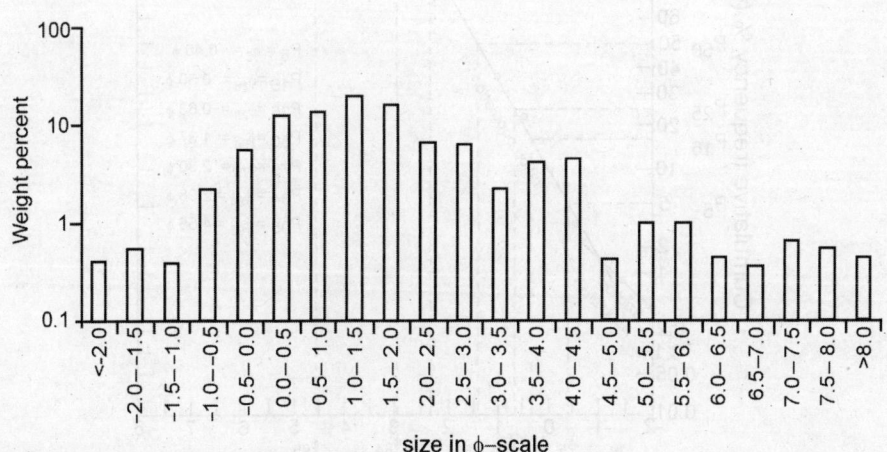

Fig. 12.2: Histogram of the grain size data of sandstone K-5 of the Ong-River Gondwana basin in logarithmic ordinate scale

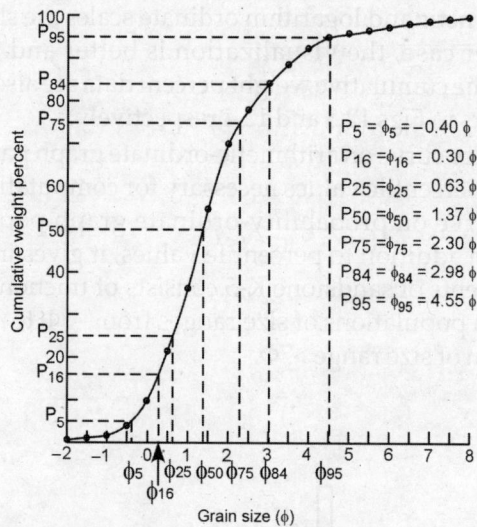

$P_5 = \phi_5 = -0.40 \phi$
$P_{16} = \phi_{16} = 0.30 \phi$
$P_{25} = \phi_{25} = 0.63 \phi$
$P_{50} = \phi_{50} = 1.37 \phi$
$P_{75} = \phi_{75} = 2.30 \phi$
$P_{84} = \phi_{84} = 2.98 \phi$
$P_{95} = \phi_{95} = 4.55 \phi$

Fig. 12.3: Graphic representation of cumulative grain-size frequency data of sandstone K-5 in arithmetic ordinate paper

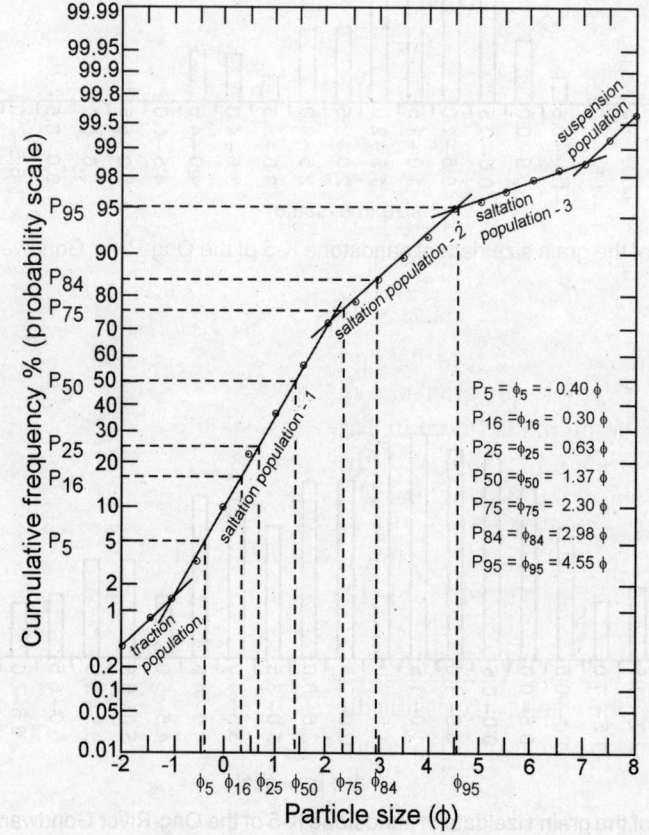

$P_5 = \phi_5 = -0.40 \phi$
$P_{16} = \phi_{16} = 0.30 \phi$
$P_{25} = \phi_{25} = 0.63 \phi$
$P_{50} = \phi_{50} = 1.37 \phi$
$P_{75} = \phi_{75} = 2.30 \phi$
$P_{84} = \phi_{84} = 2.98 \phi$
$P_{95} = \phi_{95} = 4.55 \phi$

Fig. 12.4: Graphic representation of cumulative grain-size frequency data of sandstone K-5 in probability ordinate paper

From the cumulative frequency curves, we get the following percentile values:

$$P_5 = \Phi_5 = -0.4\Phi, P_{16} = \Phi_{16} = 0.3\Phi, P_{25} = \Phi_{25} = 0.63\Phi, P_{50} = \Phi_{50} = 1.37\Phi,$$
$$P_{75} = \Phi_{75} = 2.3\Phi, P_{84} = \Phi_{84} = 2.98\Phi, \text{ and } P_{95} = \Phi_{95} = 4.55\Phi$$

Substituting these values in the formulae proposed by Folk and Ward (1957), we get

$$\text{Median} = \Phi_{50} = 1.37\Phi = 2^{-1.37} = 0.387 \text{ mm},$$

$$\text{Graphic mean (Mz)} = \frac{(\Phi_{16} + \Phi_{50} + \Phi_{84})}{3} = \frac{(0.3\Phi + 1.37\Phi + 2.98\Phi)}{3}$$

$$= \frac{4.65\ \Phi}{3} = 1.55\ \Phi = 2^{-1.55} = 0.342 \text{ mm}$$

Inclusive graphic standard deviation

$$= \frac{(\Phi_{84} - \Phi_{16})}{4} + \frac{(\Phi_{95} - \Phi_5)}{6.6}$$

$$= \frac{(2.98\Phi - 0.3\Phi)}{4} + \frac{(4.55\Phi + 0.4\Phi)}{6.6} = (0.67\Phi + 0.75\Phi)$$

$$= 1.42\Phi = 2^{-1.42} = 0.374 \text{ mm}$$

Inclusive graphic skewness

$$= \frac{(\Phi_{16} + \Phi_{84} - 2\Phi_{50})}{2(\Phi_{84} - \Phi_{16})} + \frac{(\Phi_5 + \Phi_{95} - 2\Phi_{50})}{2(\Phi_{95} - \Phi_5)}$$

$$= \frac{(0.3\Phi + 2.98\Phi - 2 \times 1.37\Phi)}{2(2.98\Phi - 0.3\Phi)} + \frac{(-0.4\Phi + 4.55\Phi - 2 \times 1.37\Phi)}{2(4.55\Phi + 0.4\Phi_5)}$$

$$= 0.101 + 0.142 = 0.243$$

$$\text{Graphic kurtosis} = \frac{(\Phi_{95} - \Phi_5)}{2.44\,(\Phi_{75} - \Phi_{25})} = \frac{(4.55\Phi + 0.4\Phi)}{2.44\,(2.3\Phi - 0.63\Phi)} = 4.95 \div 4.0748 = 1.215$$

The above parameters indicate that the sandstone K-5 is medium-grained sand, poorly sorted with fine skewed and leptokurtic grain size distribution.

Computation of moments is given in Table 12.7. The mean size (1.646Φ) and standard deviation (1.551Φ) are comparable with the corresponding values obtained from graphic measurements. However, the skewness (1.214) and kurtosis (5.122) values obtained from moment measures are higher than the corresponding values obtained from graphic measurements. This may be due to very low frequency values of the weight percent data in the intervals −2Φ to −1Φ and 4.5Φ to 8Φ (Table 12.5).

12.2 SIGNIFICANCE OF GRAIN SIZE STUDY

i. The grain size study helps in classification the sedimentary rocks of mechanical origin.

ii. The size parameters are utilized to interpret the environment as well as sub-environments of deposition.

iii. The size study helps to understand the dominant mechanism that was in operation during the deposition of sediments.

TABLE 12.7: Computation of size parameters by moment method of the Karharbari sandstone K-5

Size class (Φ)		Mid point (mi)	Wt. fr. (fi)	$f_i.m_i$	$f_i.m_i^2$	$(m_i-\mu_1)$	$(m_i-\mu_1)^3.f_i$	$(m_i-\mu_1)^4.f_i$
From	To							
-2	-1.5	-1.75	0.295	-0.516	0.903	-3.396	-11.554	39.237
-1.5	-1	-1.25	0.211	-0.264	0.330	-2.896	-5.125	14.841
-1	-0.5	-0.75	1.239	-0.929	0.697	-2.396	-17.042	40.834
-0.5	0	-0.25	3.076	-0.769	0.192	-1.896	-20.965	39.750
0	0.5	0.25	6.849	1.712	0.428	-1.396	-18.633	26.012
0.5	1	0.75	7.533	5.650	4.237	-0.896	-5.419	4.855
1	1.5	1.25	10.858	13.573	16.966	-0.396	-0.674	0.267
1.5	2	1.75	8.963	15.685	27.449	0.104	0.010	0.001
2	2.5	2.25	3.643	8.197	18.443	0.604	0.803	0.485
2.5	3	2.75	3.506	9.642	26.514	1.104	4.718	5.208
3	3.5	3.25	1.222	3.972	12.907	1.604	5.043	8.089
3.5	4	3.75	2.261	8.479	31.795	2.104	21.059	44.308
4	4.5	4.25	2.45	10.413	44.253	2.604	43.260	112.650
4.5	5	4.75	0.231	1.097	5.212	3.104	6.908	21.444
5	5.5	5.25	0.55	2.888	15.159	3.604	25.746	92.790
5.5	6	5.75	0.55	3.163	18.184	4.104	38.018	156.024
6	6.5	6.25	0.25	1.563	9.766	4.604	24.398	112.326
6.5	7	6.75	0.2	1.350	9.113	5.104	26.593	135.729
7	7.5	7.25	0.35	2.538	18.397	5.604	61.597	345.192
7.5	8	7.75	0.3	2.325	18.019	6.104	68.228	416.466
	Total		54.537	89.769	278.964		246.969	1616.508

First moment (arithmetic mean):

$$\mu_1 = \frac{\sum(f_im_i)}{\sum f_i} = 89.769\ \phi \div 54.537 = 1.646\phi$$

Second moment (variance):

$$\mu_2 = \sigma^2 = \frac{\sum(f_im_i^2) - \sum(f_im_i) \times \mu_1}{\sum f_i} = 2.406\phi^2$$

Standard deviation $(\sigma) = \sqrt{2.406\phi^2} = 1.551\phi$

Third moment (Skewness) $= \dfrac{\sum\left\{(m_i-\mu_1)^3 \times f_i\right\}}{N \times \sigma^3} = \dfrac{246.969}{(54.537 \times 3.731)} = 1.214$

Fourth moment (Kurtosis) $= \dfrac{\sum\left\{(m_i-\mu_1)^4 \times f_i\right\}}{N \times \sigma^4} = \dfrac{1616.508}{(54.537 \times 5.787)} = 5.122$

iv. It helps in studying the energy condition of the transporting medium.

v. The grain size study can be taken as an important clue to understand the sedimentation process.

vi. The sediments can be classified into different textural categories.

vii. Systematic study of grain size helps to infer palaeocurrent direction.

12.3 PROBLEMS FOR SOLUTION

The size analysis data of 12 sediment samples are given in Tables 12.8a and 12.8b. Draw the cumulative frequency curves and determine the median, mean grain size, sorting, skewness and kurtosis in each case.

TABLE 12.8a: Grain size analysis data of sediment samples

Size in phi-scale		Weight of sediment in gram					
		Q.1	Q.2	Q.3	Q.4	Q.5	Q.6
-4.0	-3.5	0.5	0.0	0.0	8.4	0.0	0.0
-3.5	-3.0	1.2	0.0	0.0	10.2	0.0	0.0
-3.0	-2.5	2.5	0.0	0.0	12	0.0	0.0
-2.5	-2.0	3.8	4.5	0.0	10.5	0.0	0.0
-2.0	-1.5	5.6	6.8	0.0	8.6	0.8	0.0
-1.5	-1.0	8.4	7.5	0.0	6.2	1.5	0.0
-1.0	-0.5	10.7	9.4	0.0	5.6	2.4	0.0
-0.5	0.0	15.7	12.8	0.0	4.3	3.7	0.0
0.0	0.5	18.4	13.6	1	3.5	5.2	0.5
0.5	1.0	20.5	10.5	2.5	2.2	7.8	0.8
1.0	1.5	12.9	9.6	4	1.3	10.4	1.2
1.5	2.0	10.6	9.2	5.3	0.6	8.3	2.4
2.0	2.5	8.4	7.4	8.4	0.2	6.4	4.6
2.5	3.0	6.2	5.7	10.6	0.1	4.6	6.6
3.0	3.5	4.5	3.8	12.8	0.0	3.5	8.7
3.5	4.0	3.1	2.5	9.3	0.0	2	10.4
4.0	4.5	0.0	1.3	7.2	0.0	1.5	10.4
4.5	5.0	0.0	0.6	6.2	0.0	0.6	8.7
5.0	5.5	0.0	4.5	3.6	0.0	0.0	6.6
5.5	6.0	0.0	0.0	2.4	0.0	0.0	4.6
6.0	6.5	0.0	0.0	1.8	0.0	0.0	2.4
6.5	7.0	0.0	0.0	0.6	0.0	0.0	1.2
7.0	7.5	0.0	0.0	0.4	0.0	0.0	0.8
7.5	8.0	0.0	0.0	0.8	0.0	0.0	0.5

TABLE 12.8b: Grain size analysis data of sediment samples

Size in phi-scale		Weight of sediment in gram					
		Q.7	Q.8	Q.9	Q.10	Q.11	Q.12
-4.0	-3.0	1.0	12.0	2.0	0.5	1.6	0.0
-3.0	-2.0	3.0	11.0	4.0	1.2	3.2	0.0
-2.0	-1.0	5.0	10.0	6.0	2.4	5.8	2.0
-1.0	0.0	7.0	9.0	8.0	4.5	12.0	4.0
0.0	1.0	9.0	8.0	10.0	6.4	10.4	6.0
1.0	2.0	12.0	7.0	12.0	8.6	8.5	8.0
2.0	3.0	8.0	6.0	14.0	9.5	6.3	6.0
3.0	4.0	6.0	5.0	16.0	12.0	4.8	4.0
4.0	5.0	4.0	4.0	18.0	8.5	2.5	2.0
5.0	6.0	3.0	3.0	20.0	6.4	1.3	0.0
6.0	7.0	2.0	2.0	22.0	3.6	0.7	0.0
7.0	8.0	1.0	1.0	24.0	1.8	0.4	0.0

12.4 ANSWERS

The cumulative frequency curves and percentile values are shown in Figs 12.5 – 12.16. The median, mean grain size, sorting, skewness and kurtosis values are presented in Table 12.9a and the conclusions are given in Table 12.9b.

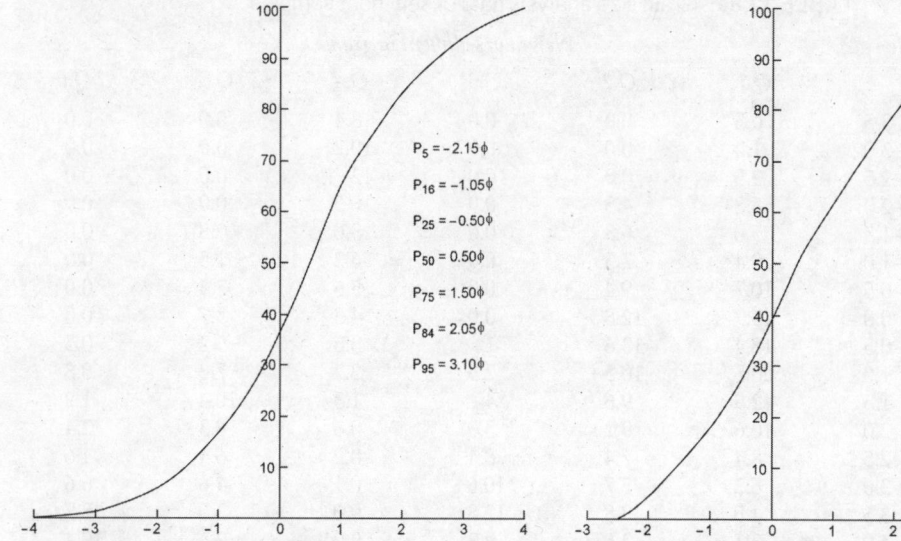

Fig. 12.5: Cumulative frequency graph of Q.1 of Table 12.8a

Fig. 12.6: Cumulative frequency graph of Q.2 of Table 12.8a

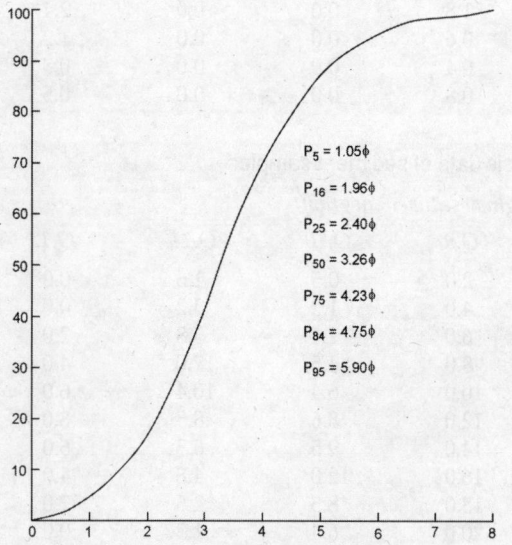

Fig. 12.7: Cumulative frequency graph of Q.3 of Table 12.8a

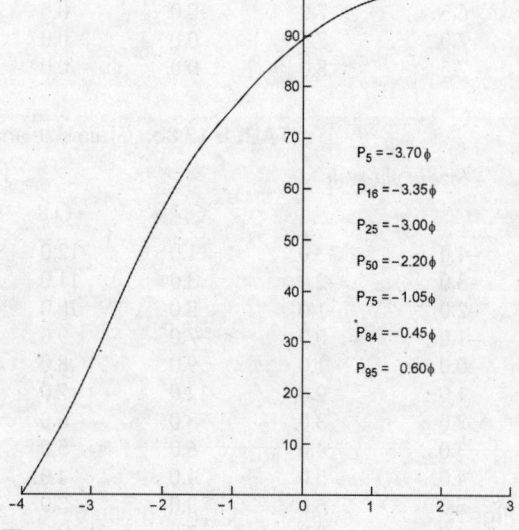

Fig. 12.8: Cumulative frequency graph of Q.4 of Table 12.8a

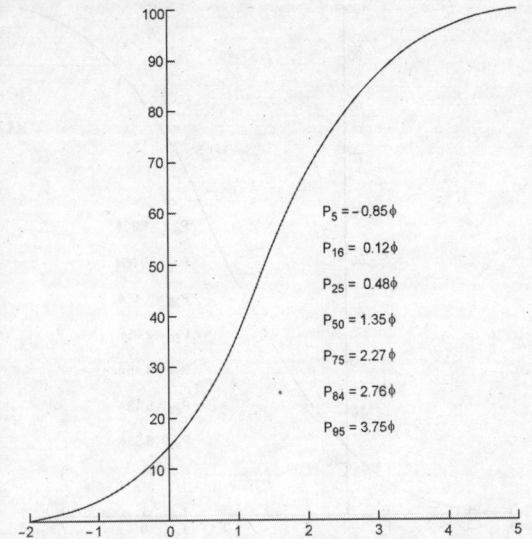

$P_5 = -0.85\phi$

$P_{16} = 0.12\phi$

$P_{25} = 0.48\phi$

$P_{50} = 1.35\phi$

$P_{75} = 2.27\phi$

$P_{84} = 2.76\phi$

$P_{95} = 3.75\phi$

Fig. 12.9: Cumulative frequency graph of Q.5 of Table 12.8a

$P_5 = 1.75\phi$

$P_{16} = 2.66\phi$

$P_{25} = 3.10\phi$

$P_{50} = 4.00\phi$

$P_{75} = 4.90\phi$

$P_{84} = 5.35\phi$

$P_{95} = 6.22\phi$

Fig. 12.10: Cumulative frequency graph of Q.6 of Table 12.8a

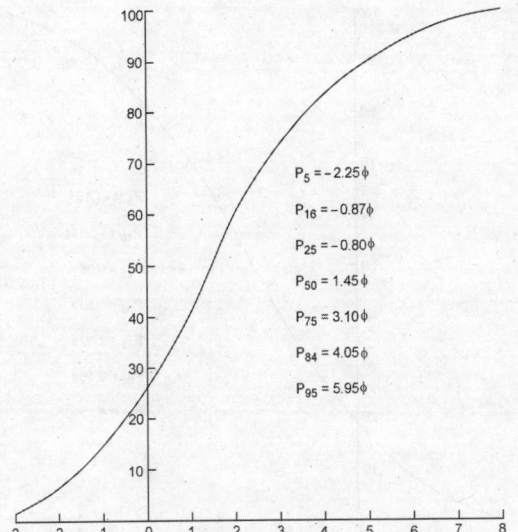

$P_5 = -2.25\phi$

$P_{16} = -0.87\phi$

$P_{25} = -0.80\phi$

$P_{50} = 1.45\phi$

$P_{75} = 3.10\phi$

$P_{84} = 4.05\phi$

$P_{95} = 5.95\phi$

Fig. 12.11: Cumulative frequency graph of Q.7 of Table 12.8b

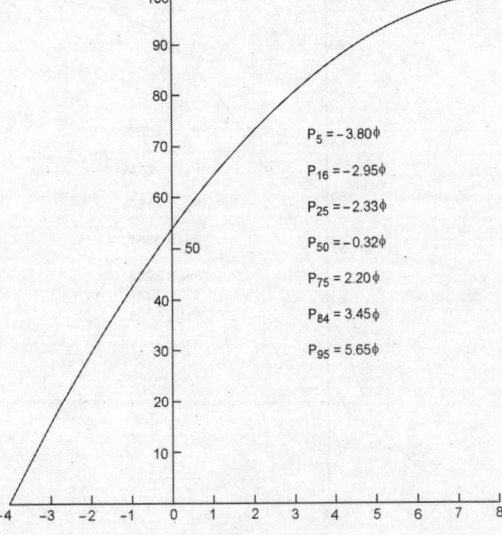

$P_5 = -3.80\phi$

$P_{16} = -2.95\phi$

$P_{25} = -2.33\phi$

$P_{50} = -0.32\phi$

$P_{75} = 2.20\phi$

$P_{84} = 3.45\phi$

$P_{95} = 5.65\phi$

Fig. 12.12: Cumulative frequency graph of Q.8 of Table 12.8b

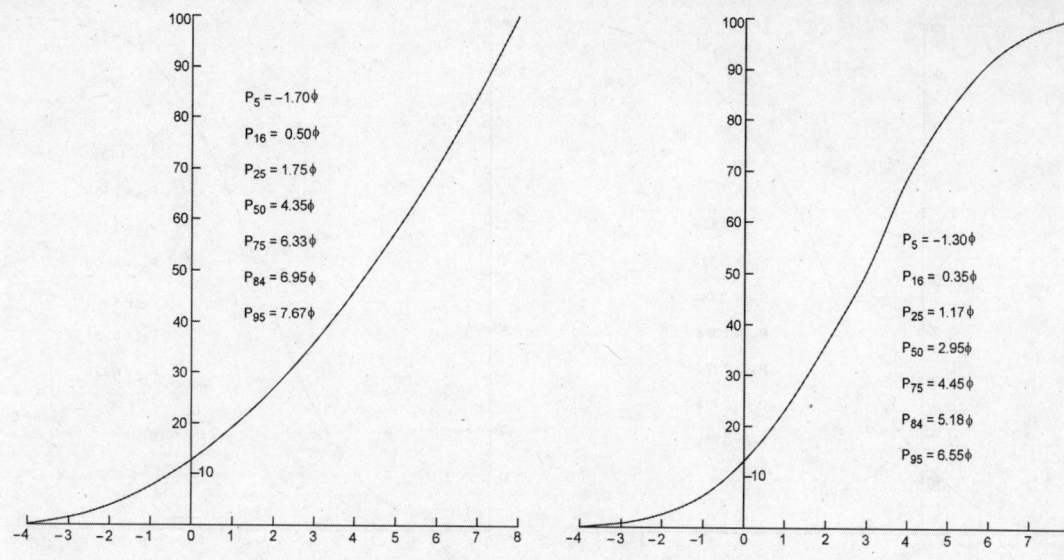

Fig. 12.13: Cumulative frequency graph of Q.9 of Table 12.8b

Fig. 12.14: Cumulative frequency graph of Q.10 of Table 12.8b

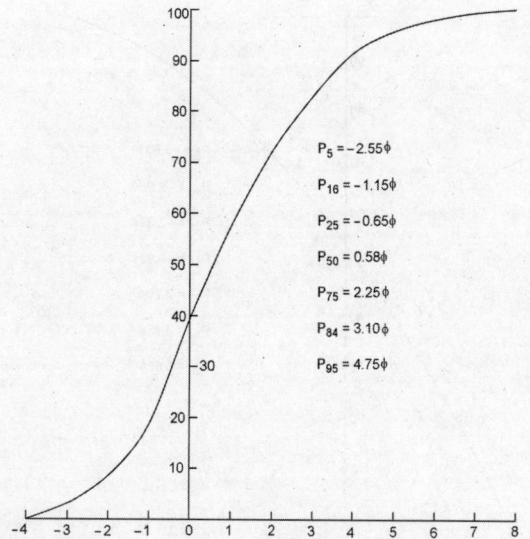

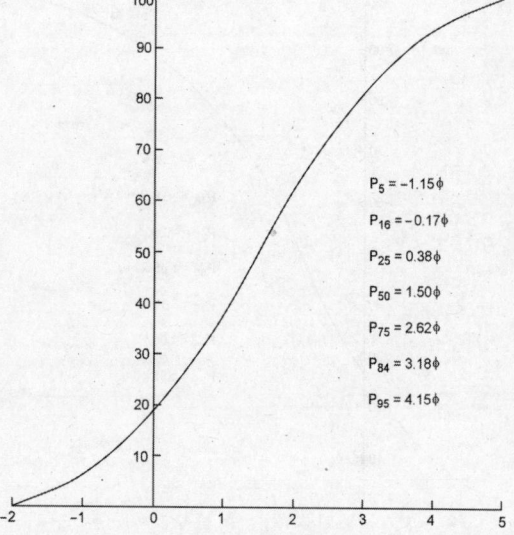

Fig. 12.15: Cumulative frequency graph of Q.11 of Table 12.8b

Fig. 12.16: Cumulative frequency graph of Q.12 of Table 12.8b

TABLE 12.9a: Grain size parameters of the problems given in Tables 12.8a and 12.8b

Q. No.	Median in Φ	Median in mm	Mean size in Φ	Mean size in mm	Standard deviation in Φ	Skewness	Kurtosis
1	0.500	0.707	0.500	0.707	1.570	−0.005	1.076
2	0.430	0.742	0.533	0.691	1.664	0.098	0.948
3	3.260	0.104	3.323	0.100	1.432	0.078	1.086
4	−2.200	4.595	−2.000	4.000	1.377	0.255	0.904
5	1.35	0.392	1.422	0.373	1.357	0.035	1.053
6	4.000	0.063	4.003	0.062	1.350	−0.001	1.018
7	1.450	0.366	1.543	0.343	2.472	0.077	0.862
8	−0.320	1.248	0.060	0.959	3.032	0.221	0.855
9	4.350	0.049	3.933	0.065	3.032	−0.243	0.838
10	2.950	0.129	2.827	0.141	2.397	−0.080	0.981
11	0.580	0.669	0.843	0.557	2.169	0.164	1.032
12	1.500	0.354	1.503	0.353	1.641	0.001	0.970

TABLE 12.9b: Conclusions of the problems given in Tables 12.8a and 12.8b

Q. No.	Conclusion
1	Coarse-grained sand, poorly sorted with nearly symmetrical and mesokurtic grain size distribution
2	Coarse-grained sand, poorly sorted with nearly symmetrical and mesokurtic grain size distribution
3	Very fine-grained sand, poorly sorted with nearly symmetrical and mesokurtic grain size distribution
4	Very fine-grained pebble, poorly sorted with fine skewed and mesokurtic grain size distribution
5	Medium-grained sand, poorly sorted with nearly symmetrical and mesokurtic grain size distribution
6	Coarse-grained silt, poorly sorted with nearly symmetrical and mesokurtic grain size distribution
7	Medium-grained sand, very poorly sorted with nearly symmetrical and platykurtic grain size distribution
8	Coarse-grained sand, very poorly sorted with fine skewed and platykurtic grain size distribution
9	Very fine-grained sand, very poorly sorted with coarse skewed and platykurtic grain size distribution
10	Fine-grained sand, very poorly sorted with nearly symmetrical and mesokurtic grain size distribution
11	Coarse-grained sand, very poorly sorted with fine skewed and mesokurtic grain size distribution
12	Medium-grained sand, poorly sorted with nearly symmetrical and mesokurtic grain size distribution

13

Palaeocurrent and Palaeohydrological Analysis

Palaeocurrent is defined as the current which has long vanished but left its imprint on rocks in form of facies, textural and structural attributes. Palaeocurrent analysis involves the study of these attributes with a view to deduce the ancient sediment dispersal pattern. There are various aspects of palaeocurrent analysis. All are based on mapping, a requirement that necessitates extended fieldwork, measurements, computations and graphic summary in map form. Palaeohydrology refers to the deduction of stream parameters like width, depth, sinuosity, water velocity, etc. of the ancient rivers which deposited the sediments in river basins. Palaeocurrent and palaeohydrological deductions can be made from the scalar and vector attributes of the sediments.

13.1 SCALAR PROPERTIES OF SEDIMENTS AND PALAEOCURRENT

The scalar properties of sediments which provide clues for deduction of palaeocurrent are variation of lithofacies, down-current decline of clast size, down-current increase of roundness and sphericity of sediments and decrease in bed thickness. Lateral variations of lithofacies are noticeable phenomena in case of sediments laid down in aqueous environments. In case of fluvial sedimentation, lateral variation of gravelly-, sandy- and shaly-facies are observed from the place of origin of river to its final destination, i.e. lake or sea. The size of pebbles and other clasts decreases in down-current direction. This declination varies widely in different situations and is the result of abrasion and other size reduction processes during transit. Roundness of gravels increases with the distance of travel. Down-current changes in the shape of gravels have also been noticed. Since both roundness and sphericity are closely correlated with size, a downstream decrease in size is commonly accompanied by an increase of roundness and sphericity. Both bedding and cross bedding show an overall decrease in thickness in the down-current direction. Systematic mapping of an area with measurement of the above mentioned parameters and their plotting in a map form can bring out the ancient sediment dispersal pattern and palaeocurrent to certain extent. Decrease of thickness of the Boulder Gravel Unit (Fig. 13.1) towards northwest suggests northwesterly palaeocurrent during Boulder Gravel sedimentation in Talcher coalfield. Change of lithofacies of the Barakar Formation from sandstone in the south to shale-coal in north (Fig. 13.2) implying overall decrease of grain size suggests northerly palaeoflow during Barakar sedimentation in Talcher coalfield.

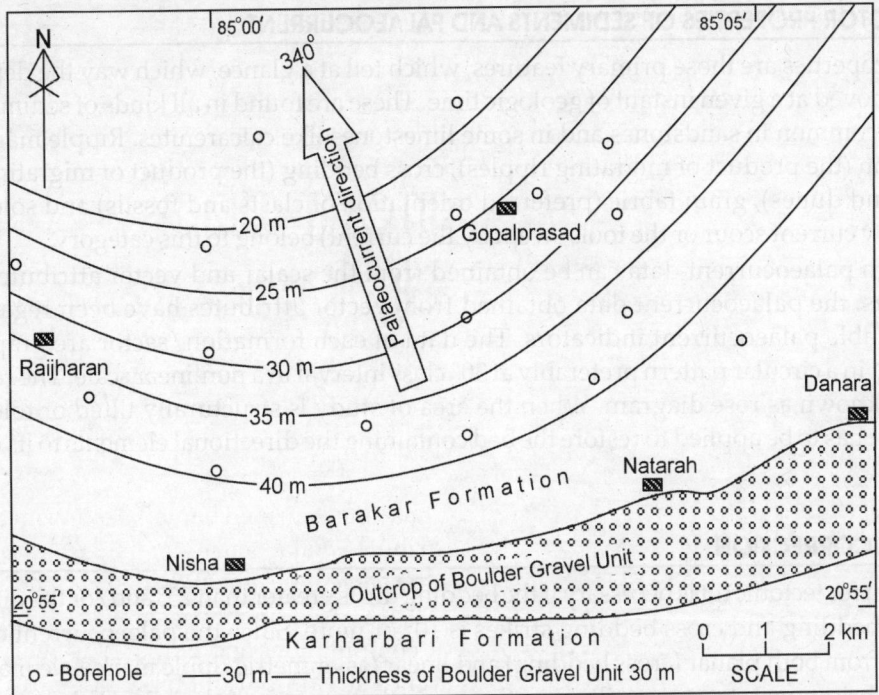

Fig. 13.1: Isopach map of the Boulder Gravel Unit of Talcher coalfield (Hota, 2007)

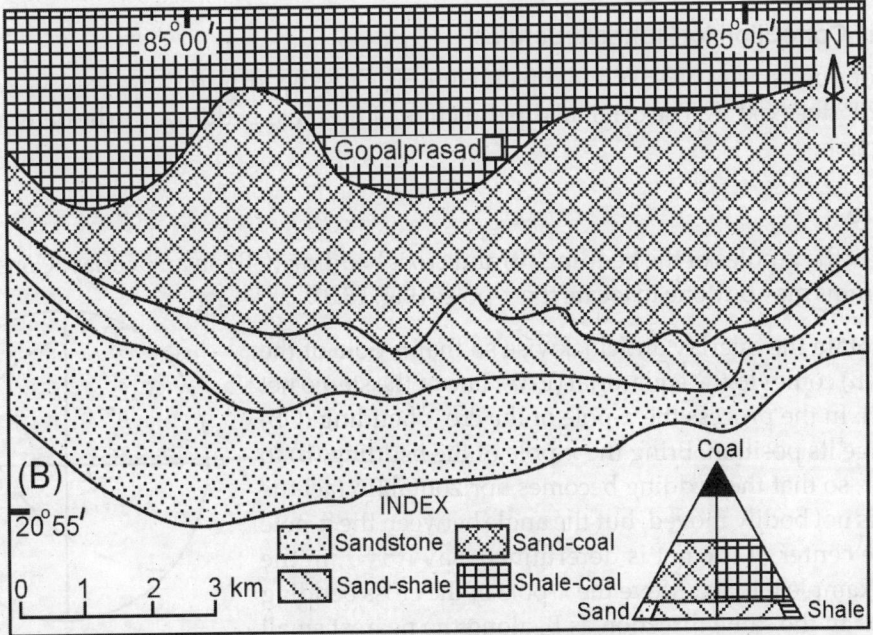

Fig. 13.2: Lithofacies map of the Barakar Formation of Talcher coalfield (Hota, 1999)

13.2 VECTOR PROPERTIES OF SEDIMENTS AND PALAEOCURRENT

Vector properties are those primary features, which tell at a glance, which way the depositing current moved at a given instant of geologic time. These are found in all kinds of sediments but are most common in sandstones and in some limestones like calcarenites. Ripple mark, cross lamination (the product of migrating ripples), cross bedding (the product of migrating mega ripples and dunes), grain fabric (preferred orientation of clasts and fossils) and sole marks (caused by current scour or the tools swept by the current) belong to this category.

Though palaeocurrent data can be obtained from the scalar and vector attributes of the sediments, the palaeocurrent data obtained from vector attributes have been regarded as more reliable palaeocurrent indicators. The data of each formation/sector are graphically presented in a circular pattern preferably at 30° class interval in a nonlinear scale. The resulting figure is known as rose diagram. When the area of study is structurally tilted or folded, tilt correction has to be applied to restore the bed containing the directional elements to its original position.

13.3 TILT CORRECTION

Correction for tectonic tilt is necessary if the bedding dip is greater than 25° and/or the difference between bedding and cross bedding strikes is 10° or more. Since the palaeocurrent data are obtained from both planar (cross bedding) and linear (asymmetric ripple mark, sole marks and pebble imbrications) structures, tilt corrections in both the cases are described below separately with examples. The correction methods involve restoration of the bedding containing the palaeocurrent structure to the horizontal position as it was at the time of deposition.

13.3.1 Tilt Correction for Planar Structures

A sandstone bed strikes NE-SW and dips 50° due SE. Cross bedding on it strike N30° W – S30° E and dip 20° due N60° E. The original dip direction of the cross bedding is to be determined by tilt correction.

Procedure

Step 1. Plot the π-poles of the bedding (B) and cross bedding (CB) on an overlay. Construction of β-diagrams may be done to locate the π-poles (Fig. 13.3).

Step 2. Rotate the overlay anticlockwise so that π-pole of the bedding (B) comes to the equatorial (E-W) line of the stereo net (Fig. 13.4). In the process the π-pole of the cross bedding (CB) will change its position. Bring the π-pole of the bedding (B) to the center, so that the bedding becomes horizontal (in fact the π-pole B is not bodily moved, but the angle between the π-pole B and the center of the net is determined; say it is α, in the present example it is 50°. Move the π-pole of the cross bedding (CB) by α° in the same direction as B, along the nearest small circle. Let the new position is CB'.

Fig. 13.3: Step 1

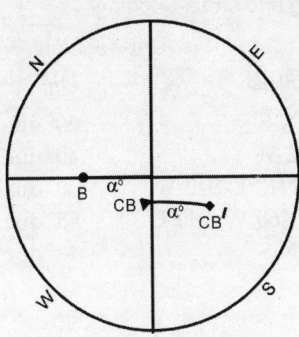

Fig. 13.4: Step 2

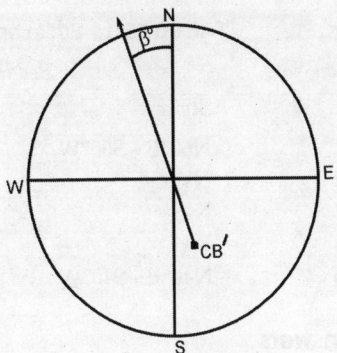

Fig. 13.5: Step 3

Step 3. Rotate the stereogram (overlay) in clockwise direction, so that the N-S line of overlay coincides with the N-S line on the stereo net (Fig. 13.5). Join CB′ and central point of the stereographic net and extend the line till it meets the opposite edge of the stereo net. The corrected cross-bedding direction is Nβ°W. In the present example it is N16°W or 344°.

13.3.2 Tilt Correction for Linear Structures

A sandstone bed strikes N50°E–S50°W and dips 50° due N40°W. The rake of an asymmetric ripple mark on the bedding plane is 40° westerly. The original azimuth of the ripple mark is to be determined by tilt correction.

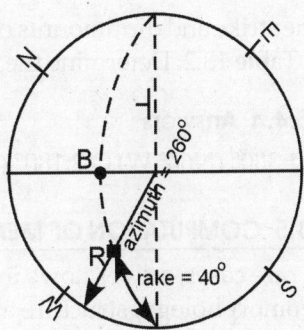

Fig. 13.6: Step 1

Step 1. Construct the β-diagram of the bedding and locate the ripple mark (R) on it by an overlay (Fig. 13.6). The azimuth of R is 260°. Let B be the intersection of the β-diagram and the E-W line of the stereo net.

Step 2. Move the point B to B′ on primitive and move the point R by the same angular distance in the same direction (Fig. 13.7) as it was done in case of planar structure described above. Let R moves to R′. Join R′ to the center of the primitive and read the changed azimuth making the N-S line of overlay coincident with the N-S line on the stereo net. In the present case it is 270°. So the corrected palaeocurrent direction is west or 270°.

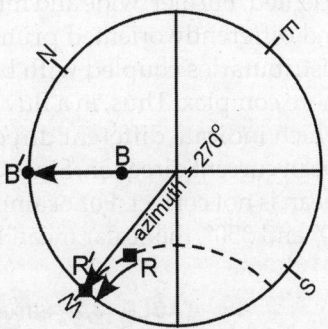

Fig. 13.7: Step 2

13.4 TILT CORRECTION PROBLEMS

13.4.1 Problems

The strike and dip amounts of bedding and cross-bedding are given in Table 13.1. Determine the original palaeocurrent directions by making corrections for tilt.

TABLE 13.1: Strike and dip values of bedding and cross-bedding

Question No.	Bedding		Cross-bedding	
	Strike	Dip	Strike	Dip
1	N60°E – S60°W	40° due N30°W	N-S	30° due W
2	NW-SE	50° due SW	E-W	40° due N
3	N-S	45° due W	N40°E–S40°W	30° due S50°E
4	E-W	40° due S	N60°W–S60°E	45° due N30°E
5	N40°E – S40°W	30° due S40°E	N-S	40° due E

13.4.2 Answers

Q.1: 210° (S30° W), Q.2: 18° (N18° E), Q.3:108° (S72° E), Q.4: 20° (N20° E) Q.5: 44° (N44°E)

13.4.3 Problems

The strike and dip amounts of bedding and plunge or rake of asymmetric ripple marks are given in Table 13.2. Determine the actual palaeocurrent directions by making corrections for tilt.

13.4.4 Answers

Q.1:300° (N60° W) Q.2:180° (S), Q.3: 296° (N64° W), Q.4: 212° (S32° W) Q.5: 60° (N60° E)

13.5 COMPUTATION OF MEAN PALAEOCURRENT DIRECTION AND OTHER STATISTICS

In rare cases, a river flows in straight line in a single direction. Rather, depending on the local geomorphology, structure and other controls, the river takes its own course. Nowadays, it is constrained to confine itself within the embankments. In geologic past, there was no human interference and the rivers were migrating according to the prevailing geological conditions. As a result, the primary structures produced by a single river at different places were differently oriented. Further, wide and multi-channel rivers are likely to have differently oriented channels and differently oriented primary sedimentary structures produced by them. Tributaries and distributaries coupled with braided and meandering nature of channels make the situation more complex. Thus, in a fluvial system, differently oriented primary structures are common, which indicate different directions at different places. Thus, it is pertinent to determine the mean current direction. Since current directions are vectors; the scalar method of estimation of mean is not correct. For example, if two current directions indicated by primary structures are 10° and 350°, the scalar mean 180°[(10° + 350°) ÷ 2] is completely ridiculous. Hence, the vector

TABLE 13.2: Strike and dip of bed and rake/plunge of asymmetric ripple mark

Question No.	Bed		Ripple mark
	Strike	Dip	
1	N60° E – S60° W	40° due N30°W	Rake 60° westerly
2	NW-SE	50° due SW	Rake 40° westerly
3	N-S	45° due W	Plunge 60° due N40 W
4	E-W	40° due S	Plunge 45° due SW
5	N40° E – S40° W	30° due S40° E	Plunge 50° due N20 E

method should be followed for computation of mean current direction. The simplest way is to plot individual vectors on a sheet of paper and determine the resultant vector.

Example: *Six palaeocurrent directions of a formation are 15°, 45°, 60°, 75°, 340° and 350°.*

Solution 1: Draw a rectangular co-ordinate OX-OY as shown in Fig. 13.8. Since all the directions are unit vectors of magnitude 1, draw 15°, 45°, 60°, 75°, 340° and 350° vectors successively of same lengths at the tip of the preceding one. Had the number of any direction more than 1, then the length of that vector would have been number of times of unit length. Finally, the tip of the last vector (350° in present case) is joined to the point 'O', which is the resultant vector (R) and its orientation with respect to 'O' is measured. In the present case it is 28°.

This method is simple and straightforward, but when the number of vectors (directions) is more, this method becomes cumbersome.

Solution 2: Each vector can be resolved into two rectangular components in OX and OY directions as shown in Fig. 13.9. The X- any Y-components can be added separately and the resultant direction $(\bar{\theta}_v)$ can be obtained from the formula $\tan \bar{\theta}_v = \dfrac{\sum Y}{\sum X}$.

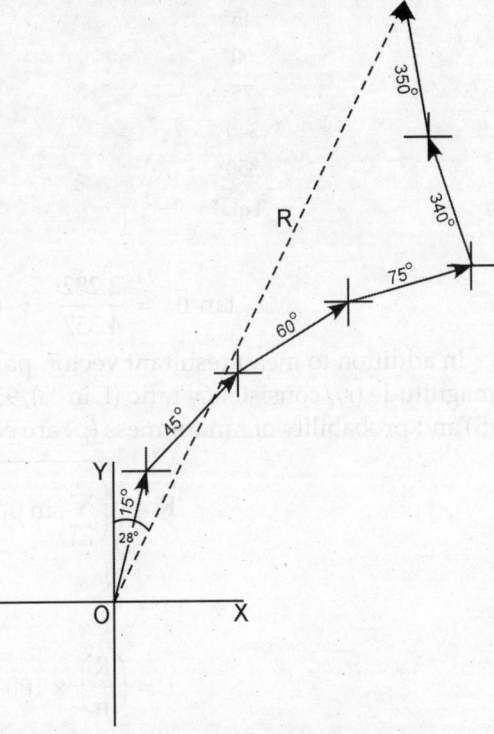

Fig. 13.8: Vector addition of palaeocurrent data

Like the previous one, this method also becomes cumbersome with a large number of data.

Solution 3: Instead of drawing individual vectors, each of them can be resolved into components in form of $\sin\theta_i$ and $\cos\theta_i$, where 'θ_i' is the azimuthal value (in degrees from north in clockwise direction) of the current direction. The sine and cosine components can be summed up separately and mean resultant vector can be determined as given below.

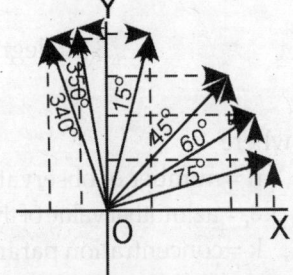

Fig. 13.9: Vector addition of palaeocurrent data by resolving into rectangular coordinates

$$\tan \bar{\theta}_v = \frac{\displaystyle\sum_{i=1}^{n} \sin \theta_i}{\displaystyle\sum_{i=1}^{n} \cos \theta_i} \Rightarrow \bar{\theta}_v = \tan^{-1}\left(\frac{\displaystyle\sum_{i=1}^{n} \sin \theta_i}{\displaystyle\sum_{i=1}^{n} \cos \theta_i}\right)$$

Let us determine the resultant vector from the above mentioned data.

Angle (θ_i)	$Sin\,\theta_i$	$Cos\,\theta_i$
15°	0.259	0.966
45°	0.707	0.707
60°	0.866	0.500
75°	0.966	0.259
340°	−0.342	0.940
350°	−0.174	0.98
Total	2.282	4.357

$$\tan \bar{\theta}_v = \frac{2.282}{4.357} \Rightarrow \bar{\theta}_v = \tan^{-1}(0.524) = 27.655° \approx 28°$$

In addition to mean resultant vector, parameters like resultant vector (R), vector strength/ magnitude (r)/consistency ratio (L in %), 95% confidence interval, circular standard deviation (S) and probability of randomness (p) are calculated by the following formulae:

$$R = \sqrt{\left(\sum_{i=1}^{n} \sin \theta_1\right)^2 + \left(\sum_{i=1}^{n} \cos \theta_1\right)^2}$$

$$r = \frac{R}{n}$$

$$L = \left(\frac{R}{n}\right) \times 100 \text{ (Expressed in percent)}$$

95% confidence interval (in degrees) $= \pm (57.3) \times 1.96 \times \dfrac{1}{\sqrt{n \times r \times k}}$

$$S \text{ (in degrees)} = (57.3) \times \sqrt{[2 \times (1 - r)]}$$

$$p = e^{-r^2 n}$$

where

n = number of observations,

θ_i = azimuthal value of the individual direction (in degrees from north in clockwise direction)

k = concentration parameter (Table 13.3).

13.5.1 Worked Out Examples

The palaeocurrent data obtained from cross-beddings of the Talchir and Karharbari Formations of the Ong-river basin and those of the Talchir Formation of the Talchir basin are given in Tables 13.4–13.6. Calculate the mean palaeocurrent direction and other statistical parameters in each case.

The computations of mean palaeocurrent and other statistical parameters of the above three formations are given in Table 13.7.

TABLE 13.3: Concentration parameter (k) values corresponding to vector magnitudes (r)

r	k	r	k	r	k	r	k	r	k
0.01	0.02000	0.21	0.42962	0.41	0.90043	0.61	1.55738	0.81	3.00020
0.02	0.04001	0.22	0.45110	0.42	0.92720	0.62	1.60044	0.82	3.14262
0.03	0.06003	0.23	0.47273	0.43	0.95440	0.63	1.64506	0.83	3.30114
0.04	0.08006	0.24	0.49453	0.44	0.98207	0.64	1.69134	0.84	3.47901
0.05	0.10013	0.25	0.51649	0.45	1.01022	0.65	1.73945	0.85	3.68041
0.06	0.12022	0.26	0.53863	0.46	1.03889	0.66	1.78953	0.86	3.91072
0.07	0.14034	0.27	0.56097	0.47	1.06810	0.67	1.84177	0.87	4.17703
0.08	0.16051	0.28	0.58350	0.48	1.09788	0.68	1.89637	0.88	4.48876
0.09	0.18073	0.29	0.60625	0.49	1.12828	0.69	1.95357	0.89	4.85871
0.10	0.20101	0.30	0.62922	0.50	1.15932	0.70	2.01363	0.90	5.30469
0.11	0.22134	0.31	0.65242	0.51	1.19105	0.71	2.07685	0.91	5.85223
0.12	0.24175	0.32	0.67587	0.52	1.22350	0.72	2.14359	0.92	6.53939
0.13	0.26223	0.33	0.69958	0.53	1.25672	0.73	2.21425	0.93	7.42572
0.14	0.28279	0.34	0.72356	0.54	1.29077	0.74	2.28930	0.94	8.61035
0.15	0.30344	0.35	0.74783	0.55	1.32570	0.75	2.36930	0.95	10.2716
0.16	0.32419	0.36	0.77241	0.56	1.36156	0.76	2.45490	0.96	12.7661
0.17	0.34503	0.37	0.79730	0.57	1.39842	0.77	2.54686	0.97	16.9266
0.18	0.36599	0.38	0.82253	0.58	1.43635	0.78	2.64613	0.98	25.2522
0.19	0.38707	0.39	0.84812	0.59	1.47543	0.79	2.75382	0.99	50.2421
0.20	0.40828	0.40	0.87408	0.60	1.51574	0.80	2.87129	1.00	∞

TABLE 13.4: Palaeocurrent data of the Talchir formation of the Ong-river basin (in degrees from north)

0	4	15	17	18	18	20	20	23	26
26	26	27	32	34	37	42	46	300	307
336	344	350	350	356	—	—	—	—	—

TABLE 13.5: Palaeocurrent data of the Karharbari formation of the Ong-river basin (in degrees from north)

12	15	24	28	33	36	157	190	198	198
198	216	230	243	246	250	252	258	260	262
265	270	270	272	273	274	274	283	286	288
292	312	316	316	318	318	323	323	325	333
334	342	347	352	358	358	—	—	—	—

TABLE 13.6: Palaeocurrent data of the Talchir Formation of the Talchir basin (in degrees from north)

2	3	3	4	5	8	13	13	23	27
33	33	46	52	53	67	70	300	305	307
308	315	318	322	328	333	334	337	338	338
340	347	347	348	353	—	—	—	—	—

TABLE 13.7: Computation of palaeocurrent parameters of Talchir and Karharbari formations of Ong-river basin and Talchir formation of Talchir basin

Ong-river basin						Talchir basin		
Talchir formation			Karharbari formation			Talchir formation		
θ_i	$Sin\ \theta$	$Cos\ \theta$	θ_i	$Sin\ \theta$	$Cos\ \theta$	θ_i	$Sin\ \theta$	$Cos\ \theta$
0	0.000	1.000	12	0.208	0.978	2	0.035	0.999
4	0.070	0.998	15	0.259	0.966	3	0.052	0.999
15	0.259	0.966	24	0.407	0.914	3	0.052	0.999
17	0.292	0.956	28	0.469	0.883	4	0.070	0.998
18	0.309	0.951	33	0.545	0.839	5	0.087	0.996
18	0.309	0.951	36	0.588	0.809	8	0.139	0.990
20	0.342	0.940	157	0.391	−0.921	13	0.225	0.974
20	0.342	0.940	190	−0.174	−0.985	13	0.225	0.974
23	0.391	0.921	198	−0.309	−0.951	23	0.391	0.921
26	0.438	0.899	198	−0.309	−0.951	27	0.454	0.891
26	0.438	0.899	198	−0.309	−0.951	33	0.545	0.839
26	0.438	0.899	216	−0.588	−0.809	33	0.545	0.839
27	0.454	0.891	230	−0.766	−0.643	46	0.719	0.695
32	0.530	0.848	243	−0.891	−0.454	52	0.788	0.616
34	0.559	0.829	246	−0.914	−0.407	53	0.799	0.602
37	0.602	0.799	250	−0.940	−0.342	67	0.921	0.391
42	0.669	0.743	252	−0.951	−0.309	70	0.940	0.342
46	0.719	0.695	258	−0.978	−0.208	300	−0.866	0.500
300	−0.866	0.500	260	−0.985	−0.174	305	−0.819	0.574
307	−0.799	0.602	262	−0.990	−0.139	307	−0.799	0.602
336	−0.407	0.914	265	−0.996	−0.087	308	−0.788	0.616
344	−0.276	0.961	270	−1.000	0.000	315	−0.707	0.707
350	−0.174	0.985	270	−1.000	0.000	318	−0.669	0.743
350	−0.174	0.985	272	−0.999	0.035	322	−0.616	0.788
356	−0.070	0.998	273	−0.999	0.052	328	−0.530	0.848
Total	4.398	22.067	274	−0.998	0.070	333	−0.454	0.891
			274	−0.998	0.070	334	−0.438	0.899
			283	−0.974	0.225	337	−0.391	0.921
			286	−0.961	0.276	338	−0.375	0.927
			288	−0.951	0.309	338	−0.375	0.927
			292	−0.927	0.375	340	−0.342	0.940
			312	−0.743	0.669	347	−0.225	0.974
			316	−0.695	0.719	347	−0.225	0.974
			316	−0.695	0.719	348	−0.208	0.978
			318	−0.669	0.743	353	−0.122	0.993
			318	−0.669	0.743	Total	−1.962	28.864
			323	−0.602	0.799			
			323	−0.602	0.799			
			325	−0.574	0.819			

(Contd...)

TABLE 13.7: Computation of palaeocurrent parameters of Talchir and Karharbari formations of Ong-river basin and Talchir formation of Talchir basin (*Contd...*)

	Ong-river basin				Talchir basin
	Talchir formation		*Karharbari formation*			*Talchir formation*
θ_i		θ_i	*Sin* θ	*Cos* θ	θ_i
		333	−0.454	0.891	
		334	−0.438	0.899	
		342	−0.309	0.951	
		347	−0.225	0.974	
		352	−0.139	0.990	
		358	−0.035	0.999	
		358	−0.035	0.999	
		Total	**−23.923**	**11.184**	
$\bar{\theta}_v$	11°		295°		356°
n	25		46		35
R	22.501		26.4082		28.9306
r	0.9000		0.5741		0.8266
k	5.30469		1.39842		3.30114
L	90.00%		57.41%		82.66%
95% con. int.	± 10°		± 18°		± 11°
cir. std. dev. (S)	24°		52°		33°
p	$< 10^{-8}$		$< 10^{-6}$		$< 10^{-10}$

13.6 GRAPHIC REPRESENTATION OF PALAEOCURRENT DATA

The palaeocurrent data is graphically represented in the form of circular histogram known as rose diagram. In case of rectangular histogram, the area of each bar is proportional to the corresponding frequency (the widths of each bar are kept constant and thus the height of each bar is made proportional to the corresponding frequency). The width of each bar is selected depending on the lowest and highest values of data. The palaeocurrent data varies in the circular scale varying from 0° to 360° . Hence the class intervals can be made 10°, 15°, 20°, 30°, 45°and 90°. 10° interval makes the number of class too many (36), whereas 90° makes the number of class too less (4). Commonly rose diagrams are constructed at interval of 30°. Since the rose diagram is a circular histogram, the area of each petal (sector) should be proportional to the frequency of corresponding sector.

$$\text{Area of a sector (A)} = \frac{1}{2} r^2 \theta \ (r \text{ is the radius of the sector and } \theta \text{ is the sector angle in radian})$$

$$= \frac{90}{\pi} r^2 \theta \ (\theta \text{ is the degree}) = \frac{2700}{\pi} r^2 \ (\text{if } \theta \text{ is } 30°)$$

$$A = K r^2 \text{ or } A \propto r^2 \ (K \text{ is a constant})$$

If for 1% frequency radius (r) is made 1 cm, then A = 1 K cm² (any value other than 1 cm can also be taken; 1 cm has been taken for simplicity)

In linear scale, for 2% frequency, $r = 2$ cm then $A = K(2)^2 = 4K$ cm^2

for 3% frequency, $r = 3$ cm and $A = K(3)^2 = 9K$ cm^2

for 4% frequency, $r = 4$ cm and $A = K(4)^2 = 16K$ cm^2 so on

i.e. in linear scale, the area of sector varies with square of the frequency.

In non-linear scale, for f% frequency if we take $r = \sqrt{f}$

for 2% frequency, $r = \sqrt{2}$ cm then $A = K(\sqrt{2})^2 = 2K$ cm^2

for 3% frequency, $r = \sqrt{3}$ cm and $A = K(\sqrt{3})^2 = 3K$ cm^2

for 4% frequency, $r = \sqrt{4}$ cm and $A = K(\sqrt{4})^2 = 4K$ cm^2 so on

i.e in non-linear scale (for f% frequency if we take $r = \sqrt{f}$), the area of each sector is proportional to the corresponding frequency.

In such a scale for 1% frequency, $r = \sqrt{1}$ cm and $A = K(\sqrt{1})^2 = 1K$ cm^2

for 5% frequency, $r = \sqrt{5}$ cm and $A = K(\sqrt{5})^2 = 5K$ cm^2

for 10% frequency, $r = \sqrt{10}$ cm and $A = K(\sqrt{10})^2 = 10K$ cm^2

for 20% frequency, $r = \sqrt{20}$ cm and $A = K(\sqrt{20})^2 = 20K$ cm^2 so on

Hence, the rose diagrams should always be drawn in non-linear scale taking radius of sector equal to square root of frequency.

The computations of radii for construction of rose diagrams of the data given in Tables 13.4– 13.6 are given in Table 13.8 and the corresponding rose diagrams are presented in Figs 13.10– 13.12.

TABLE 13.8: Computation of radii of sectors of rose diagrams

| Class interval (in degrees) | Ong-river basin | | | | | | Talchir basin | | |
| | Talchir formation | | | Karharbari formation | | | Talchir formation | | |
	f	f%	r	f	f%	r	f	f%	r
0–30	13	52	7.2	4	8.70	2.9	10	28.57	5.3
30–60	5	20	4.5	2	4.35	2.1	5	14.29	3.8
60–90	0	0	0	0	0.00	0.0	2	5.71	2.4
90–120	0	0	0	0	0.00	0.0	0	0.00	0.0
120–150	0	0	0	0	0.00	0.0	0	0.00	0.0
150–180	0	0	0	1	2.17	1.5	0	0.00	0.0
180–210	0	0	0	4	8.70	2.9	0	0.00	0.0
210–240	0	0	0	2	4.35	2.1	0	0.00	0.0
240–270	0	0	0	8	17.39	4.2	0	0.00	0.0
270–300	0	0	0	10	21.74	4.7	0	0.00	0.0
300–330	2	8	2.8	8	17.39	4.2	8	22.86	4.8
330–360	5	20	4.5	7	15.22	3.9	10	28.57	5.3
n =	25			46			35		

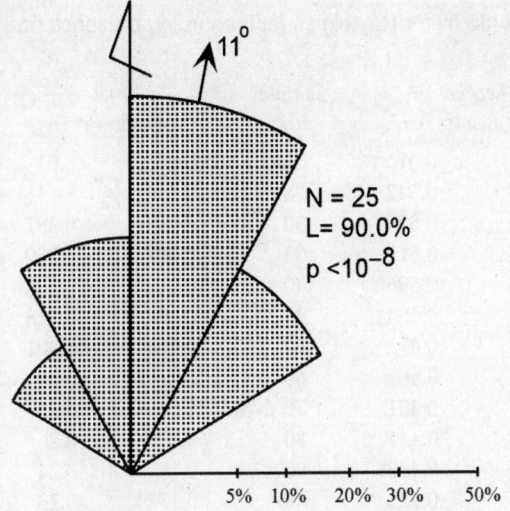

Fig. 13.10: Rose diagram of Talchir formation of Ong- river basin

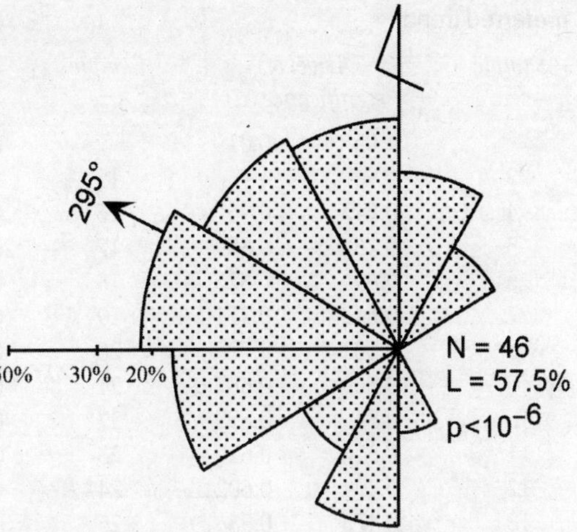

Fig. 13.11: Rose diagram of Karharbari formation of Ong- river basin

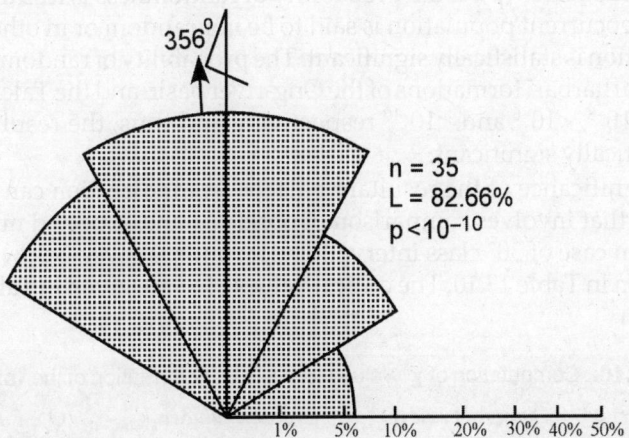

Fig. 13.12: Rose diagram of Talchir formation of Talchir basin

13.7 STATISTICAL SIGNIFICANCE OF RESULTANT PALAEOCURRENT DIRECTION

The statistical significance of the resultant palaeocurrent direction can be quickly determined by Rayleigh's test that involves comparison of the computed value of vector strength / magnitude (r) with that of critical value given in Table 13.9. In cases of the Talchir and Karharbari formations of the Ong-river basin and the Talchir formation of the Talchir basin, the sample sizes, i.e. number of data are 25, 46 and 35 respectively for which the critical values are 0.423, ~0.315 and 0.359 respectively. Since the calculated values of 'r' (0.9, 0.5741 and 0.8266 respectively) are much higher than the corresponding critical values at 0.01 significance level, the resultant palaeocurrent directions are statistically significant.

TABLE 13.9: Critical values of vector strength / magnitude (r) for Rayleigh's test regarding presence of a preferred trend

Sample size	Level of significance (α)		Sample size	Level of significance (α)		Sample size	Level of significance (α)	
	0.05	0.01		0.05	0.01		0.05	0.01
3	0.96	—	15	0.443	0.542	28	0.33	—
4	0.847	0.96	16	0.429	0.525	30	0.315	0.387
5	0.754	0.879	17	0.417	0.51	35	0.292	0.359
6	0.69	0.825	18	0.405	0.596	40	0.273	0.336
7	0.642	0.771	19	0.394	0.484	45	0.257	0.318
8	0.602	0.725	20	0.385	0.472	50	0.244	0.301
9	0.569	0.687	21	0.375	0.461	60	0.22	0.30
10	0.54	0.655	22	0.367	0.451	70	0.21	0.27
11	0.516	0.627	23	0.359	0.441	80	0.19	0.26
12	0.494	0.602	24	0.351	0.432	90	0.18	0.24
13	0.475	0.58	25	0.344	0.423	100	0.17	0.23
14	0.458	0.56	26	0.34	—	200	0.12	0.16

The statistical significance of the mean palaeocurrent direction can also be examined in terms of probability of randomness (p). If the probability of randomness is less than 0.01 ($<10^{-2}$), the distribution of palaeocurrent population is said to be nonrandom or in other words, the mean palaeocurrent direction is statistically significant. The probability of randomness values in cases of the Talchir and Karharbari formations of the Ong-river basin and the Talchir formation of the Talchir basin are $<10^{-8}$, $<10^{-6}$ and $<10^{-10}$ respectively and thus, the resultant palaeocurrent directions are statistically significant.

The statistical significance of the resultant palaeocurrent direction can also be verified by Chi-square (χ^2) test that involves comparison of observed and expected number of data in all 12 possible sectors in case of 30° class interval. The formula is given below and the method of computation is given in Table 13.10. The critical values of χ^2 are given in Table 13.11, which are used for comparison.

TABLE 13.10: Computation of χ^2 values for the Talchir formation of the Talchir basin

Class interval (in degrees)	Observed value (O_j)	Expected value (E_j)	$(O_j - E_j)^2$	$[(O_j - E_j)^2 / E_j]$
0–30	10	2.92	50.13	17.17
30–60	5	2.92	4.33	1.48
60–90	2	2.92	0.85	0.29
90–120	0	2.92	8.53	2.92
120–150	0	2.92	8.53	2.92
150–180	0	2.92	8.53	2.92
180–210	0	2.92	8.53	2.92
210–240	0	2.92	8.53	2.92
240–270	0	2.92	8.53	2.92
270–300	0	2.92	8.53	2.92
300–330	8	2.92	25.81	8.84
330–360	10	2.92	50.13	17.17
Total	35	35.04	190.92	$\chi^2 = 65.38$

TABLE 13.11: Critical values of χ^2

Degrees of freedom (v)	Significance level (α)/confidence level (%)					
	0.001 (99.9%)	0.005 (99.5%)	0.01 (99%)	0.025 (97.5%)	0.05 (95%)	0.1 (90%)
1	10.83	7.879	6.635	5.024	3.841	2.706
2	13.82	10.597	9.210	7.378	5.991	4.605
3	16.27	12.838	11.345	9.348	7.815	6.251
4	18.47	14.860	13.277	11.143	9.488	7.779
5	20.52	16.750	15.086	12.832	11.070	9.236
6	22.46	18.548	16.812	14.449	12.892	10.645
7	24.32	20.278	18.475	16.013	14.067	12.017
8	26.12	21.955	20.090	17.535	15.507	13.362
9	27.88	23.589	21.666	19.023	16.919	14.684
10	29.59	25.188	23.209	20.483	18.307	15.987
11	31.26	26.757	24.725	21.920	19.675	17.275
12	32.91	28.300	26.217	23.337	21.026	18.549
13	34.53	29.819	27.688	24.736	22.362	19.812
14	36.12	31.319	29.141	26.119	23.685	21.064
15	37.70	32.801	30.578	27.488	24.996	22.307
16	39.25	34.267	32.000	28.845	26.296	23.542
17	40.79	35.718	33.409	30.191	27.587	24.769
18	42.31	37.156	34.805	31.526	28.869	25.989
19	43.82	38.582	36.191	32.852	30.144	27.204
20	45.31	39.997	37.566	34.170	31.410	28.412
21	46.80	41.40	38.93	35.48	32.670	29.62
22	48.27	42.796	40.289	36.781	33.924	30.813
23	49.73	44.18	41.640	38.08	35.170	32.01
24	51.18	45.558	42.980	39.364	36.415	33.196
25	52.62	46.93	44.310	40.65	37.650	34.38
26	54.05	48.290	45.642	41.923	38.885	35.563
27	55.48	49.64	46.960	43.19	40.110	36.74
28	56.89	50.994	48.278	44.461	41.337	37.916
29	58.30	52.34	49.590	45.72	42.560	39.09
30	59.70	53.672	50.892	46.979	43.773	40.256
40	73.40	66.766	63.690	59.342	55.758	51.805
50	86.66	79.490	76.154	71.420	67.505	63.167
60	99.61	91.952	88.379	83.298	79.082	74.397
70	112.3	104.215	100.425	95.023	90.531	85.527
80	124.8	116.321	112.329	106.629	101.879	96.578
90	137.2	128.299	124.116	118.136	113.145	107.565
100	149.4	140.170	135.807	129.561	124.342	118.498
110			147.410	140.92	135.480	129.39
120	173.6	163.6	158.950	152.21	146.57	140.23
200	267.5	255.264	249.445	241.058	233.994	226.021

$$\chi^2 = \frac{\sum\limits_{j=1}^{12}(O_j - E_j)^2}{E_j}$$

Where

E_j = expected number of data in class j; (j varies from 1 to 12 for 30° class interval)

O_j = observed number of data in class j; (j varies from 1 to 12 for 30° class interval)

If the computed value of χ^2 is less than the critical value (19.675) for 11 degree of freedom at 5% significance level (Table 13.11), then the null hypothesis, i.e. palaeocurrent data are uniformly distributed is accepted. In such case, the rose diagram will not show any preferential distribution of palaeocurrent directions. On the other hand, if the computed value of χ^2 is greater than 19.675, the null hypothesis is rejected. In this case, the rose diagram will show a significant preferential palaeocurrent direction. It is to be noted that in the present χ^2 test, the degrees of freedom is 11, i.e. the number of sectors (n = 12 in case of 30° class interval) minus 1. The computed value of χ^2 in case of Talchir formation of the Talchir basin is 65.38 which is greater than the critical value of 19.675. This suggests statistical significance of the resultant palaeocurrent direction. The χ^2 values for the Talchir and Karharbari formations of the Ong-river basin are 82.17 and 36.99 respectively.

13.8 COMPARISON OF PALAEOCURRENT POPULATIONS OF TWO FORMATIONS

When the analysis of palaeocurrent of more than one formation is involved, it is customary to compare equality of the palaeocurrent populations. For this, the data of pairs of formations are pooled together and F-test is performed to test the equality of palaeocurrent populations by the following test static:

$$F = \left(1 + \frac{3}{8k}\right)\left(\frac{(N-2)(R_A + R_B + R_T)}{(N - R_A - R_B)}\right)$$

where

N = total number of data of formations A and B

R_A = resultant vector of formation A

R_B = resultant vector of formation B

R_T = pooled vector resultant of formations A and B

Example 1: *Comparison of palaeocurrent data of (A) Talchir formation of Ong-river basin and (B) Talchir formation of Talchir basin*

$$N = 25 + 35 = 60, R_A = 22.501, R_B = 28.9306$$
$$R_T = \{(4.398 - 1.962)^2 + (22.067 + 28.864)^2\}^{\frac{1}{2}} = 50.989$$
$$\Rightarrow \qquad r = 50.989 \div 60 = 0.85$$

For, r = 0.85, k = 3.68041

Computed value of F = 3.296 < $F_{\alpha = 5\%, (1,59)}$ (approximately 4.00) (Table 13.12)

⇒ Palaeocurrent samples came from populations with the same mean direction

⇒ Palaeocurrent populations of Talchir formation of Ong-river basin and Talchir formation of Talchir basin are equal.

TABLE 13.12: Critical values of 'F' at 0.05 significance level / 95% confidence level													
$v_1 \rightarrow$ $v_2 \downarrow$	1	2	3	4	5	6	7	8	9	10	15	20	25	∞
1	161.45	199.50	215.71	224.58	230.16	233.99	236.77	238.88	240.54	241.88	245.95	248.01	249.26	250.10
2	18.51	19.00	19.16	19.25	19.30	19.33	19.35	19.37	19.38	19.40	19.43	19.45	19.46	19.46
3	10.13	9.55	9.28	9.12	9.01	8.94	8.89	8.85	8.81	8.79	8.70	8.66	8.63	8.62
4	7.71	6.94	6.59	6.39	6.26	6.16	6.09	6.04	6.00	5.96	5.86	5.80	5.77	5.75
5	6.61	5.79	5.41	5.16	5.05	4.95	4.88	4.82	4.77	4.74	4.62	4.56	4.52	4.50
6	5.99	5.14	4.76	4.53	4.39	4.28	4.21	4.15	4.10	4.06	3.94	3.87	3.83	3.81
7	5.59	4.74	4.35	4.12	3.97	3.87	3.79	3.73	3.68	3.64	3.51	3.44	3.40	3.38
8	5.32	4.46	4.07	3.84	3.69	3.58	3.50	3.44	3.39	3.35	3.22	3.15	3.11	3.08
9	5.12	4.26	3.86	3.63	3.48	3.37	3.29	3.23	3.18	3.14	3.01	2.94	2.89	2.86
10	4.96	4.10	3.71	3.48	3.33	3.22	3.14	3.07	3.02	2.98	2.85	2.77	2.73	2.70
11	4.84	3.98	3.59	3.36	3.20	3.90	3.01	2.59	2.90	2.85	2.72	2.65	2.60	2.57
12	4.75	3.89	3.49	3.26	3.11	3.00	2.91	2.85	2.80	2.75	2.62	2.54	2.50	2.47
13	4.67	3.81	3.41	3.18	3.03	2.92	2.83	2.77	2.71	2.67	2.53	2.46	2.41	2.38
14	4.60	3.74	3.34	3.11	2.96	2.85	2.76	2.70	2.65	2.60	2.46	2.39	2.34	2.31
15	4.54	3.68	3.29	3.06	2.90	2.79	2.71	2.64	2.59	2.54	2.40	2.33	2.28	2.25
16	4.49	3.63	3.24	3.01	2.85	2.74	2.66	2.59	2.54	2.49	2.35	2.28	2.23	2.19
17	4.45	3.59	3.20	2.96	2.81	2.70	2.61	2.55	2.49	2.45	2.31	2.23	2.18	2.15
18	4.41	3.55	3.16	2.93	2.77	2.66	2.58	2.51	2.46	2.41	2.27	2.19	2.14	2.11
19	4.38	3.52	3.13	2.90	2.74	2.63	2.54	2.48	2.42	2.38	2.23	2.16	2.11	2.07
20	4.35	3.49	3.10	2.87	2.71	2.60	2.51	2.45	2.39	2.35	2.20	2.12	2.07	2.04
21	4.32	3.47	3.07	2.84	2.68	2.57	2.49	2.42	2.37	2.32	2.18	2.10	2.05	2.01
22	4.30	3.44	3.05	2.82	2.66	2.55	2.46	2.40	2.34	2.30	2.15	2.07	2.02	1.98
23	4.28	3.42	3.03	2.80	2.64	2.53	2.44	2.37	2.32	2.27	2.13	2.05	2.00	1.96
24	4.26	3.40	3.01	2.78	2.62	2.51	2.42	2.36	2.30	2.25	2.11	2.03	1.97	1.94
25	4.24	3.39	2.99	2.76	2.60	2.49	2.40	2.34	2.28	2.24	2.09	2.01	1.96	1.92
26	4.23	3.37	2.98	2.74	2.59	2.47	2.39	2.32	2.27	2.22	2.07	1.99	1.94	1.90
27	4.21	3.35	2.96	2.73	2.57	2.46	2.37	2.31	2.25	2.20	2.06	1.97	1.92	1.88
28	4.20	3.34	2.95	2.71	2.56	2.45	2.36	2.29	2.24	2.19	2.04	1.96	1.91	1.87
29	4.18	3.33	2.93	2.70	2.55	2.43	2.35	2.28	2.22	2.18	2.03	1.94	1.89	1.85
30	4.17	3.32	2.92	2.69	2.53	2.42	2.33	2.27	2.21	2.16	2.01	1.93	1.88	1.84
40	4.08	3.23	2.84	2.61	2.45	2.34	2.25	2.18	2.12	2.08	1.92	1.84	1.78	1.74
50	4.03	3.18	2.79	2.56	2.40	2.29	2.20	2.13	2.07	2.03	1.87	1.78	1.73	1.69
60	4.00	3.15	2.76	2.53	2.37	2.25	2.17	2.10	2.04	1.99	1.84	1.75	1.69	1.65
70	3.98	3.13	2.74	2.50	2.35	2.23	2.14	2.07	2.02	1.97	1.81	1.72	1.66	1.62
80	3.96	3.11	2.72	2.49	2.33	2.21	2.13	2.06	2.00	1.95	1.79	1.70	1.64	1.60
90	3.95	3.10	2.71	2.47	2.32	2.20	2.11	2.04	1.99	1.94	1.78	1.69	1.63	1.59
100	3.94	3.09	2.70	2.46	2.31	2.19	2.10	2.03	1.97	1.93	1.77	1.68	1.62	1.57
110	3.93	3.08	2.69	2.45	2.30	2.18	2.09	2.02	1.97	1.92	1.76	1.67	1.61	1.56
120	3.92	3.07	2.68	2.45	2.29	2.18	2.09	2.02	1.96	1.91	1.75	1.66	1.60	1.55
∞	3.85	3.00	2.61	2.38	2.22	2.11	2.02	1.95	1.89	1.84	1.68	1.58	1.52	1.47

Example 2: *Comparison of palaeocurrent data of (A) Talchir formation of Ong-river basin and (B) Karharbari formation of Ong-river basin*

$$N = 25 + 46 = 71, R_A = 22.501, R_B = 26.4082$$
$$RT = \{(4.398 - 23.923)^2 + (22.067 + 11.184)^2\}^{1/2} = 38.56,$$
$$\Rightarrow \qquad r = 38.56 \div 71 = 0.54$$

For r = 0.54, k = 1.29077

Computed value of F = 41.7 > $F_{\alpha=5\%, (1,70)}$ (3.98) (Table 13.12)

⇒ Palaeocurrent samples came from populations with the different mean directions

⇒ Palaeocurrent populations of Talchir and Karharbari formations of Ong-river basin are significantly different from each other.

13.9 PALAEOCURRENT PROBLEMS FOR SOLUTION

13.9.1 The dip directions of fore-sets of tabular cross beddings of 12 stratigraphic units are given in Table 13.13. Determine the mean palaeocurrent direction and other statistical parameters in each case. Draw the rose diagrams in non-linear scale.

13.9.2 The answers are given in Table 13.14 and rose diagrams are shown in Fig. 13.13.

TABLE 13.13: Palaeocurrent data (The figures are degrees from north in clockwise direction)

Q-1	2	3	3	4	8	13	13	23	27	33
	33	67	70	300	305	307	308	315	318	322
	328	333	334	337	338	340	347	347	348	353
Q-2	0	0	0	0	5	5	10	12	13	13
	15	17	17	18	20	24	25	32	36	37
	42	43	46	47	53	54	63	74	88	93
	95	102								
Q-3	277	295	295	296	297	308	312	320	324	328
	333	338								
Q-4	6	7	12	16	300	312	313	318	320	325
	325	327	327	327	328	328	332	332	333	333
	338	342	342	343	348	352	357			
Q-5	2	3	7	8	12	13	18	18	20	22
	23	25	26	28	28	32	35	43	45	52
	324	325	326	326	333	340	342	343	346	346
	346	348	352	353	357	357	357	358	358	
Q-6	2	3	3	4	4	7	8	8	10	17
	18	26	37	43	298	322	324	327	332	338
	343	345	351	351	352	353	353	355	355	357
	357	358								
Q-7	0	0	17	23	27	42	260	283	287	292
	300	303	303	304	307	310	313	314	320	322
	324	327	328	330	330	333	350	353	355	357
Q-8	254	270	273	275	276	277	277	277	278	279
	280	282	282	282	283	283	283	284	285	286
	286	288	288	288	292	296	298	300	300	303
	306	307	317							

TABLE 13.13: Palaeocurrent data (The figures are degrees from north in clock-wise direction) (*Contd.*)

Q-9	8	23	237	240	245	253	257	260	262	264
	265	270	274	276	276	277	277	282	283	284
	285	285	286	288	291	292	292	295	296	296
	297	297	302	304	313	313	313	315	316	317
	317	318	320	325	327	327	332	334	335	
Q-10	12	294	297	302	305	314	320	324	324	324
	326	328	333	337	347	356				
Q-11	262	266	267	270	277	277	280	282	283	285
	286	293	294	300	300	300	302	303	304	304
	304	307	313	315	315	317	318	324	327	333
	333	336	343	350						
Q-12	13	17	18	35	38	43	300	301	307	315
	318	333	336	343	349	350				

TABLE: 13.14: Answers of the palaeocurrent problems

Q. No.	n	Vector mean	R	r	L	SD	95% con. int.	p	Chi sq.
1	30	351°	25.3433	0.8448	84.48 %	32°	12°	$<10^{-8}$	63.600
2	32	33°	27.9546	0.8736	87.36 %	29°	10°	$<10^{-10}$	113.50
3	12	310°	11.4237	0.9520	95.20%	18°	10°	$<10^{-4}$	28.250
4	27	336°	25.6071	0.9484	94.84 %	18°	7°	$<10^{-10}$	83.375
5	39	364°	35.8924	0.9203	92.03 %	23°	7°	$<10^{-14}$	138.125
6	32	356°	29.9210	0.9350	93.50 %	21°	7°	$<10^{-14}$	110.750
7	30	329°	25.3693	0.8456	84.56 %	32°	12°	$<10^{-9}$	67.250
8	33	286°	32.2614	0.9776	97.76 %	12°	4°	$<10^{-13}$	233.375
9	49	295°	42.7894	0.8733	87.33 %	29°	8°	$<10^{-16}$	202.125
10	16	325°	15.0110	0.9382	93.82 %	20°	10°	$<10^{-6}$	38.250
11	34	302°	31.4135	0.9239	92.39 %	22°	8°	$<10^{-12}$	110.250
12	16	348°	13.2954	0.8310	83.10 %	33°	17°	$<10^{-4}$	25.500

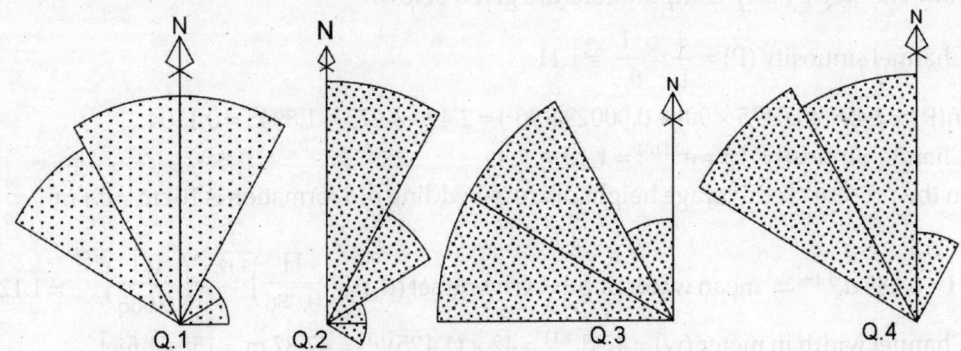

Fig. 13.13: Rose diagrams (*Contd...*)

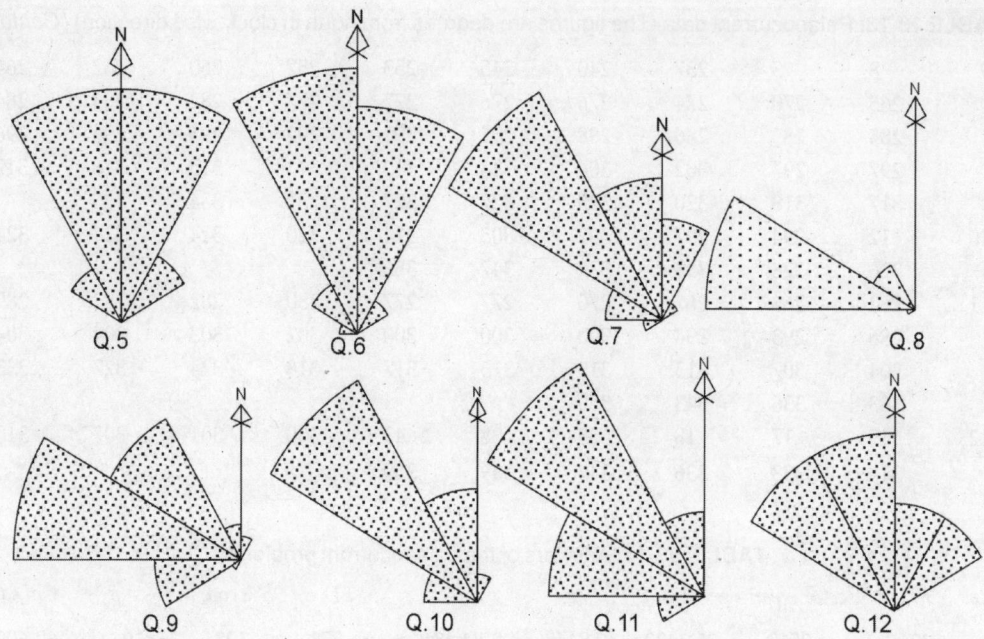

Fig. 13.13: Rose diagrams

13.10 PALAEOHYDROLOGICAL ANALYSIS

Palaeohydrologic analysis refers to estimation of channel parameters like sinuosity (P), mean (d_s) and bank full (d_b) water depths, width (w), width/depth ratio (F), sediment load parameter (M), meander wavelength (L_m), mean (Q_m) and flood (Q_{ma}) discharge, slope (S_c), normal (v) and flood stage (v_f) velocities and Froude number (F_r) from a set of empirical equations given by different researchers. These equations are given in Table 13.15.

Example: *The average height of cross-bedding of a formation is 10 cm and the vector magnitude is 0.9. The palaeohydrologic parameters of the ancient stream are to be computed.*

Solution: The step by step computations are given below:

i. Channel sinuosity $(P) = \dfrac{1}{L} = \dfrac{1}{0.9} = 1.11$

ii. $\ln(P) = 2.49 - (0.0475 \times 90) + 0.000234(90^2) = 2.49 - 4.275 + 1.8945 = 0.1104$

    Channel sinuosity $(P) = e^{0.1104} = 1.12$

iii. In the present case average height of cross-bedding of a formation is 10 cm = 0.1 m

$$H = 0.086\,d_s^{1.19} \Rightarrow \text{mean water depth in river tract } (d_s) = \left(\frac{H}{0.086}\right)^{\frac{1}{1.19}} = \left(\frac{0.1}{0.086}\right)^{0.84} = 1.125\,\text{m}$$

iv. Channel width in meter (w) = $42\,d_s^{1.11} = 42 \times (1.125)^{1.11} = 47.87\,\text{m} = 158.61\,\text{feet}$

v. Width/depth ratio (F) = 47.87 m ÷ 1.125 = 42.55

TABLE 13.15: Formulae employed for computation of palaeohydrologic parameters (Hota *et al.*, 2007)

Formula:

Channel sinuosity (P) = 1/L

Channel sinuosity (P) = 0.94 $M^{0.25}$

ln(P) = 2.49 − 0.0475L + 0.000234$(L)^2$ (L in percent)

Mean cross bed thickness in meter (H) = 0.086 $d_s^{1.19}$

Channel width in meter (w) = 42 $d_s^{1.11}$

Width/depth ratio (F) = 225 $M^{-1.08}$

Meander wavelength in feet (L_m) = 10.9 $w^{1.01}$

Meander wavelength in feet (L_m) = 106.1 $Q_m^{0.46}$

Channel slope in feet/mile (S_c) = 60$M^{-0.38}$ $Q_m^{-0.32}$

Bank full depth in feet (d_b) = 0.6 $M^{0.34}Q_m^{0.29}$

Bank full depth in feet (d_b) = 0.09 $M^{0.35}Q_{ma}^{0.42}$

Flow velocity in m/sec (v) = Q_m/wd_s

Flood stage velocity in m/sec (v_f) = Q_{ma}/wd_b

Froude number (F_r) = $v/(gd_s)^{1/2}$.

Explanation of symbols:

L = vector magnitude/ vector strength/consistency ratio

M = sediment load parameter = percentage of silt and clay in stream channel perimeter

d_s = mean water depth in river tract in meter

Q_m = mean annual discharge in feet^3/sec

Q_{ma} = mean annual flood discharge in feet^3/sec

g = acceleration due to gravity (9.8 m/sec)

vi. Width/depth ratio (F) = 225 $M^{-1.08}$ $\Rightarrow$ M = $\left(\dfrac{F}{225}\right)^{-\frac{1}{1.08}}$ = $\left(\dfrac{42.55}{225}\right)^{-0.926}$ = 4.67

vii. Meander wavelength (L_m) in feet = 10.9 $w^{1.01}$ = 10.9 × $(158.61)^{1.01}$ = 1818.70 feet = 554.34 m

viii. Meander wavelength in feet (L_m) = 106.1 $Q_m^{0.46}$

$\Rightarrow$ Mean annual discharge (Q_m) in feet^3/sec = $\left(\dfrac{L_m}{106.1}\right)^{\frac{1}{0.46}}$ = $\left(\dfrac{1818.70}{106.1}\right)^{2.174}$ = 481.74 feet^3/sec

= 13.64 m^3/sec

ix. Channel slope in feet/mile (S_c) = 60$M^{-0.38}Q_m^{-0.32}$ = 60 × $(4.67)^{-0.38}$ × $(481.74)^{-0.32}$ = 4.63 ft/mile

= 87.66 cm/km ≈ 0.00088

x. Bank full depth in feet (d_b) = 0.6 $M^{0.34}Q_m^{0.29}$ = 0.6 × $(4.67)^{0.34}$ × $(481.74)^{0.29}$ = 6.077 feet = 1.85 m

xi. Bank full depth in feet (d_b) = 0.09 $M^{0.35}Q_{ma}^{0.42}$

$\Rightarrow$ Mean annual flood discharge (Q_m) in feet^3/sec =

= $\left(\dfrac{6.077}{0.09 \times 4.67^{0.35}}\right)^{\frac{1}{0.42}}$ = $(39.3716)^{2.381}$ = 6282.63 = 177.81 m^3/sec

xii. Flow velocity (v) in m/sec = $\dfrac{Q_m}{w \times d_s}$ = $\dfrac{13.64}{47.87 \times 1.125}$ = 0.25 m/sec = 250 cm/sec

xiii. Flood stage velocity (v_f) in m/sec = $\dfrac{Q_m}{w \times d_b}$ = $\dfrac{177.81}{47.87 \times 1.85}$ = 2.0 m/sec

xiv. Froude number (F_r) = $\dfrac{v}{\sqrt{g \times d_s}}$ = $\dfrac{0.25}{\sqrt{9.8 \times 1.125}}$ = 0.075.

Conclusion

The palaeostreams were low sinuous (1.11–1.12), shallow (1.125–1.85 m) with average width of 47.87 m and meander wavelength of 554.34 m. They flowed over a gently sloping surface of 88 cm/km with velocities ranging from 0.25 to 2.0 m/sec. The hydrodynamic character (Froude number = 0.075) indicates that the stream flow was in the lower flow regime that gave rise to small-scale bed configurations. The discharge was of the order of 13.64–177.81 m^3/sec.

13.10.1 Problems for Solution

The average height of cross-bedding and vector magnitude of some formations are given in Table 13.16. Determine the palaeohydrologic parameters of the ancient streams.

13.10.2 Answers

The palaeohydrologic parameters of the ancient streams given in Table 13.16 are presented in Table 13.17.

TABLE 13.16: Cross-bedding thickness and vector magnitude of some formations

Sl. No.	Formation	Average cross-bedding thickness (H) in cm	Vector magnitude (r)
1	Karharbari formation of Talchir basin	38	0.7436
2	Barakar formation of Talchir basin	19	0.7317
3	Barren measures formation of Talchir basin	25	0.6381
4	Talchir formation of Ong-river basin	30	0.9000
5	Karharbari formation of Ong-river basin	25	0.5741
6	Barakar formation of Korba coal field	40	0.7754

TABLE 13.17: Estimates of palaeohydrologic parameters of palaeostreams

*Formation/parameters	1	2	3	4	5	6
Channel sinuosity (P)	1.33	1.36	1.48	1.21	1.17	1.27
Mean water depth (d_s)	3.49 m	1.95 m	2.45 m	2.86 m	2.45 m	3.64m
Bankfull water depth (d_b)	3.94 m	2.66 m	3.11 m	3.45 m	3.11 m	4.05m
Channel width (w)	168 m	88 m	114 m	135 m	114 m	176m
Width/depth ratio (F)	48.14	45.13	46.53	47.14	46.35	48.41
Sediment load parameter (M)	4.17	4.43	4.30	4.25	4.32	4.15
Meander wavelength (L_m)	1951 m	1015 m	1318 m	1561 m	1314 m	2046 m
Mean annual discharge (Q_m)	210 m^3/sec	51 m^3/sec	90 m^3/sec	129 m^3/sec	89 m^3/sec	233 m^3/sec
Mean annual flood discharge (Q_{ma})	1190 m^3/sec	442 m^3/sec	654 m^3/sec	843 m^3/sec	651 m^3/sec	1266 m^3/sec
Channel slope (S_c)	0.00038	0.00059	0.00050	0.00044	.00050	0.00037
Flow velocity (v)	0.358 m/s	0.297 m/s	0.322 m/s	0.336 m/s	0.320 m/s	0.36m/s
Flood stage velocity (v_f)	1.80 m/s	1.88 m/s	1.84 m/s	1.82 m/s	1.84 m/s	1.77m/s
Froude number (F_r)	0.061	0.068	0.066	0.064	0.065	0.061

*Formations: 1. Karharbari formation of Talchir basin, 2. Barakar formation of Talchir basin, 3. Barren measures formation of Talchir basin, 4. Talchir formation of Ong-river basin, 5. Karharbari formation of Ong-river basin, 6. Barakar formation of Korba coal field

13.11 SIGNIFICANCE OF PALAEOCURRENT ANALYSIS

i. Palaeocurrent indicates the position of the source area with respect to the basin of deposition.

ii. In fluvial environment, palaeocurrent is slope controlled and as such indicates the palaeoslope of the drainage basin.

iii. Palaeocurrent indicates the dominant direction of current competency.

iv. In mechanical concentration deposits, the palaeocurrent helps in ascertaining the position of mother load.

v. Different statistical attributes of the palaeocurrent help in discriminating different types of environments.

vi. Palaeocurrent analysis serves as a tool for basin analysis.

vii. The palaeocurrent analysis helps to determine the geometry of the sedimentary bodies. An important byproduct of palaeocurrent analysis is the establishment of relation between palaeocurrent and shape and configurations of individual sand bodies and also of carbonate reefs and distribution of debris derived from them.

viii. Palaeocurrent is helpful in ascertaining sedimentary strike.

ix. It helps in inferring the trend of the shore line, and also the trend and location of the margin of the depositional basin (assists in exploring the relation between the palaeocurrent system and basin architecture or geometry).

x. Stability of palaeocurrent systems implies the existence of stable tectonic elements with their consequent slopes that govern erosion, transportation and sedimentation for long periods of time.

xi. Palaeocurrent analysis has greatest value in palaeogeographic reconstruction, as there exists a distinct relationship between the physical geography of the seas, estuaries and rivers and the currents present in them. Currents impress themselves on the deposits formed under their influence and their characters could be ascertained from those formed in the ancient times.

14

Heavy Mineral Analysis

The accessory minerals present in sedimentary rocks with specific gravity greater than 2.87 (specific gravity of bromoform) are termed heavy minerals. Their total concentration rarely exceeds 1% of the total volume of the rock. The frequency of heavy minerals shows inverse relationship with the grain size. The heavy minerals are commonly collected from sandstones of grain size 0.125 to 0.088 mm (120–170 ASTM mesh number). Though sands coarser than 0.125 mm contain some heavy mineral, its concentration is less and bigger grain size creates problem for proper identification. Similarly, sands finer than 0.088 mm, though contain plenty heavy minerals, finer grain size creates problem for proper identification. However, collection of heavy minerals from beach sands up to 0.074 mm size is recommended. The type and quantity of heavy minerals present in a sedimentary rock are dependent upon a number of factors such as:

 i. Composition of the source rock
 ii. Stability of the heavy mineral
 iii. Nature of the depositional environment
 iv. Action of intrastratal solution
 v. Climate and weathering
 vi. Depth of burial
 vii. Distance and duration of transportation
viii. Age of the rock

Instead of a single mineral, a group of heavy minerals known as heavy mineral suite are taken into consideration for study of source rock lithology. Different rocks have their characteristic heavy mineral assemblages. These are given in Table 14.1.

14.1 METHODS OF HEAVY MINERAL SEPARATION

Since the heavy minerals constitute about 1% of the total volume of the sedimentary rock, special procedures are adopted to separate them from their associated light minerals. These processes are:

TABLE 14.1: Characteristic heavy mineral suits of different types of rocks

Rock type		Characteristic heavy minerals
Reworked sediments	:	Barite, iron ores, rounded rutile, tourmaline and zircon
Low-rank metamorphic	:	Biotite, chlorite, glaucophane, brown tourmaline (euhedral)
High-rank metamorphic	:	Andalucite, chloritoid, diopside, epidote, garnet, blue-green hornblende, kyanite, sillimanite, staurolite
Granite pegmatite	:	Fluorite, garnet, muscovite, topaz, monazite, blue tourmaline (indicolite)
Acid igneous rock	:	Apatite, biotite, brownish green hornblende, monazite, muscovite, rutile, sphene, pink euhedral tourmaline, euhedral zircon
Basic igneous rock	:	Anatase, augite, diopside, enstatite, hypersthene, reddish brown hornblende, olivine, magnetite, chromite

14.1.1 Hand Picking Method

In this method, the sample from which heavy minerals are to be separated is evenly spread over a glass slide and viewed under a binocular microscope with high magnification. The heavy minerals are identified by their characteristic properties and picked up by a pin.

14.1.2 Magnetic Method

This method is based on the difference in magnetic susceptibility of different minerals. Some of the minerals are highly magnetic while others are completely nonmagnetic. On the basis of magnetic susceptibility the minerals can be divided into four groups. These are given in Table 14.2.

Separation of the heavy minerals is done with the help of hand magnet and electromagnet. The hand magnet is effective only for strongly magnetic minerals while less magnetic minerals are separated conveniently by electromagnet. The electromagnet consists of a 'U' shaped soft iron covered by insulated copper wires (Fig. 14.1). Two poles of the magnet are fixed by two adjustable plates, which can be set at different distances ranging from 1.0–0.2 cm. The insulated copper wires are connected to the two poles of a dry battery. When electric current flows through the copper wires, the poles of the soft iron are magnetized. The intensity of the magnetism can be varied by adjusting the distance between magnetic poles. With the help of hand magnet, highly magnetic minerals are separated. Placing the magnetic poles 1.0 cm apart, moderately magnetic minerals are picked up. The weakly magnetic minerals are separated by keeping the magnetic poles closer, about 0.5 cm or less apart.

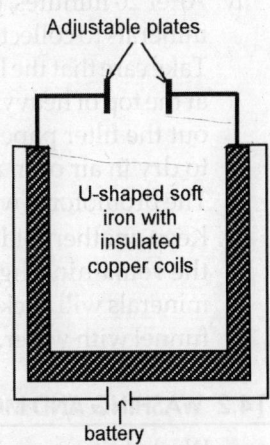

Adjustable plates

U-shaped soft iron with insulated copper coils

battery

Fig. 14.1: Electromagnet

TABLE 14.2: Magnetic properties of common minerals

Magnetic property		Minerals
Strongly magnetic	:	Magnetite, pyrrhotite
Moderately magnetic	:	Hematite, ilmenite, chromite, sphalerite, cordierite, garnet, etc.
Weakly magnetic	:	Monazite, chlorite, diopside, epidote, olivine, tourmaline, etc.
Nonmagnetic	:	Quartz, feldspar, etc.

14.1.3 Electrical Method

This method is based on the electric conducting capacity of the minerals. On the basis of electric conducting capacity, the minerals are grouped under three categories. These are given in Table 14.3. Electrostatic and dielectric separation methods are commonly employed for separation of heavy minerals.

14.1.4 Heavy Liquid Method

This is the most convenient method for separation of heavy minerals. The pieces of equipment for heavy mineral separation are normally available in all the well-stocked laboratories. In this procedure, the sandstone is disaggregated and 0.125–0.088 mm (or any other specified) size fraction is separated by sieving prior to heavy mineral analysis.

The procedure is outlined below:

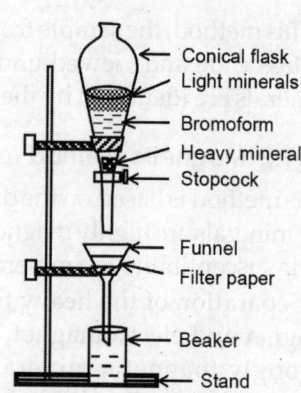

   i. Setup the apparatuses as shown in Fig. 14.2.

   ii. Pour the heavy liquid (bromoform) into the conical flask. Be sure that the stopcock is securely closed and the beaker is in its place. Put the sand sample containing the heavy minerals into the flask and stir with a glass rod.

   iii. Allow the minerals to settle down for about 20 minutes and stir the bromoform with the glass rod again.

   iv. After 20 minutes, open the stopcock and allow the heavy minerals to collect on the filter paper placed in the funnel. Take care that the liquid containing light minerals floating at the top of heavy liquid do not pass into the funnel. Take out the filter paper with the heavy minerals and allow it to dry in air or in an oven at temperature less than 50°C. The bromoform will evaporate.

Fig. 14.2: Laboratory set up of instruments for heavy mineral separation

   v. Keep another folded filter paper on the funnel and open the stopcock completely so that the remaining light fraction along with bromoform drains down. Some of the light minerals will stick to the lower part of the funnel, which can be separated by washing the funnel with water.

14.2 WASHING AND MOUNTING OF HEAVY MINERALS

   i. Wash the heavy minerals by acetone and $SnCl_2$ to remove bromoform and iron coating and dry.

   ii. Weigh the heavy fraction and calculate the weight percentage of the heavy fraction.

TABLE 14.3: Electrical properties of common heavy minerals		
Electric property		*Minerals*
Good conductor	:	Graphite, magnetite, hematite, pyrrhotite, pyrite, ilmenite, rutile, etc.
Moderate conductor	:	Iron rich pyroxenes and amphiboles, anatase, biotite, cassiterite, tourmaline, etc.
Poor conductors	:	Zircon, spinel, corundum, apatite, etc.

iii. Remove the magnetic fraction with a hand magnet and calculate the weight percent of magnetic fraction. If possible, they are to be studied in reflected polarized light and/or X-ray diffraction analysis.

iv. Put a drop of water on a glass slide and pour about 1000 heavy minerals selected by coning and quartering. Heat the slide on a hot plate. The water will evaporate and the heavy minerals will stick to the slide.

v. Pour a drop of canada balsam onto the slide and heat on the hot plate. The fluidity of canada balsam will increase and it will spread over the heavy fraction.

vi. Put a clean cover glass and rotate it so that the heavy minerals will spread evenly and air bubbles come out.

vii. Cool the slide and clean it with xylene.

viii. Identify the heavy minerals by their optical properties under a petrological microscope using refracted polarizing light. Calculate the number frequency of each species by field count or ribbon count method. Since the individual opaque heavy minerals cannot be identified in refracted polarizing light, group the heavy minerals as *opaque* and *non-opaque* fractions. The opaque minerals can be mounted in a suitable medium and polished by special technique and studied under petrological microscope using reflected polarizing light.

14.3 IDENTIFICATION OF HEAVY MINERALS

After the slide is ready, the next step is the identification of heavy minerals. During transportation, the corners of individual minerals are broken and the grains are rounded to certain extent. The shape and optical characters are distorted to some extent. The opaque grains appear black and cannot be conclusively identified in refracted polarized light. They are to be mounted in a medium, polished and identified in reflected polarized light. However, ilmenite invariably alters to leucoxene and occasionally unaltered core of black ilmenite occurs within dead white leucoxene. Siderite, which at times occurs in parting sandstones of coal bearing sediments, is identified by brass-yellow colour, rhombohedral, spherulitic or colloform structure. The synopsis of identification criteria of common heavy minerals is given in Fig. 14.3 and diagnostic characters of some transparent heavy minerals are given in Table 14.4.

14.4 COUNTING OF HEAVY MINERALS

The presence or absence of a particular heavy mineral is much more important than its size and frequency of occurrence. The conventional methods (area, line and point counting) employed in modal analysis are not applicable in heavy mineral counting. This is apparent from Fig. 14.4 in which different sizes of euhedral zircons are shown. If the area of crystal A is 1 unit, then the area of crystal B is 4 units. In conventional modal analysis, the unit at B should be counted as equivalent to 4 crystals, which is not true. The lengths of lines intercepted by crystals C, D and E are one, two and three units respectively, even though each one is a single crystal. In ordinary line count method, these should be one, two and three crystals respectively. A case may arise where the entire crystal is not represented in the traverse line, as in case of crystal F. In point count method, the cross wire may jump over a crystal (G), where the mineral is not included in

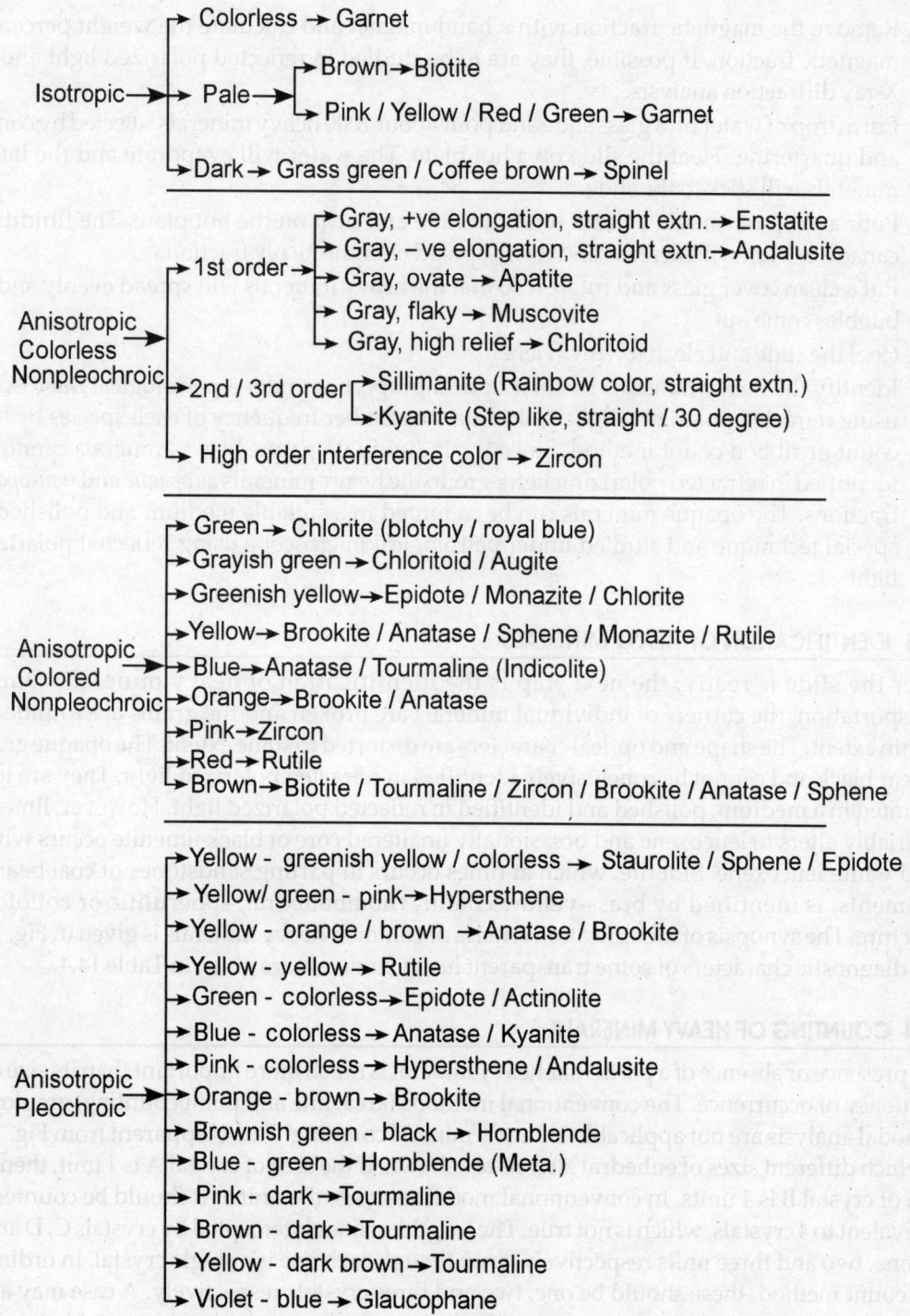

Fig. 14.3: Synopsis of identification criteria of common heavy minerals

TABLE 14.4: Optical properties of some common transparent heavy minerals

	Optical properties	Actinolite	Anatase	Andalusite	Apatite	Augite
Plane polarized light	Pleochroism / colour	Pleochroic, colourless to pale yellowish green or brown	Weakly pleochroic, yellow, brown, deep blue, orange	Nonpleochroic, or weakly pleochroic from colourless to pale pink	Nonpleochroic, colourless	Nonpleochroic, brownish gray or pale grayish green
	Refractive index	High	High	High	High	High
	Shape and habit	Fibrous	Tabular, rectangular with striations, prismatic, geometrical partitioning	Anahedral or elongated with conchoidal fracture	Rounded, oval, short prismatic	Subhedral, prismatic
	Cleavage	Imperfect	Imperfect	Imperfect	Imperfect	2 sets perfect
	Alteration	Commonly to talc	Uncommon	Sillimanite or sericite	Uncommon	Common, cloudy due to alteration
	Inclusion	Uncommon	Uncommon	Common	Inclusion in rows	Common
Crossed nicol	Optic character	Biaxial –	Uniaxial –	Biaxial –	Uniaxial –	Biaxial +
	Interference colour	Second order	Second to third order	First order gray	First order gray	Second order
	Birefringence	Moderate	Moderate	Low	Low	Moderate
	Extinction	10°–20°	Parallel	Parallel	Parallel	40°–45°
	Twinning	Polysynthetic	Absent	Absent	Absent	Polysynthetic
	Sign of elongation	Positive	—	Negative	Negative	—
	Optic axial angle	79°–85°	—	80°–85°	—	58°–62°

(Contd...)

TABLE 14.4: Optical properties of some common transparent heavy minerals (*Contd...*)

	Optical properties	Barite	Biotite	Brookite	Cassiterite
Plane polarized light	Pleochroism/colour	Nonpleochroic, colourless	Nonpleochroic in basal section, brown, occasionally green	Weakly pleochroic, orange, brown, yellow	Pleochroic, colourless to yellow to brown (blood red)
	Refractive index	High	High	Very high	High
	Shape and habit	Tabular, irregular	Equant, tabular, basal sections common	Squarish with truncated corners, striations parallel to long axis	Prism terminated with low pyramid, striations on prism face, conchoidal fracture
	Cleavage	Imperfect	1 set perfect	Not seen	Imperfect
	Alteration	Uncommon	Common	Uncommon	Uncommon
	Inclusion	Numerous	Pleochroic halos	Uncommon	Uncommon
Crossed nicol	Optic character	Biaxial +	Biaxial –		
	Basal sections are isotropic	Biaxial +	—		
	Interference colour	First order yellow	Second order (often masked by body colour)	High order brown	—
	Birefringence	Low	High, low in basal section	High	—
	Extinction	Symmetrical prismatic grains—parallel	Parallel	Incomplete due to high dispersion	—
	Twinning	Polysynthetic	—	Absent	—
	Sign of elongation	—	—	—	—
	Optic axial angle	36°–37°	Low	0°–30°	—

(Contd...)

TABLE 14.4: Optical properties of some common transparent heavy minerals (*Contd...*)

Optical properties		Chlorite	Chloritoid	Corundum	Diopside
Plane polarized light	Pleochroism/colour	Weakly pleochroic, pale green to yellowish green	Pleochroic, colourless to greenish gray	Nonpleochroic, colourless, pink; occasionally pleochroic from blue to bluish green	Nonpleochroic colourless to pale green
	Refractive index	Low to moderate	Moderately high	High	Moderate
	Shape and habit	Equant, rounded, irregular, micaceous	Equant, irregular	Irregular	Subhedral, prismatic, rounded
	Cleavage	1 set	1 set perfect	Parting present	2 sets perfect
	Alteration	Common	Common	Uncommon	Chlorite, cloudy due to alteration
	Inclusion	Uncommon	Present	Uncommon	Common
Crossed nicol	Optic character	Biaxial –	Biaxial +	Uniaxial –	Biaxial +
	Interference colour	Abnormal colours (royal blue), blotchy	First order colours	Second order	Second order colours
	Birefringence	Low to moderate	Moderate	Low	High
	Extinction	—	0–20	Parallel	Symmetrical
	Twinning	Absent	Polysynthetic	Lamellar	Polysynthetic
	Sign of elongation	—	—	Positive	Positive
	Optic axial angle	0°–30°	45° – 68°	—	56°–64°

(Contd...)

TABLE 14.4: Optical properties of some common transparent heavy minerals (*Contd...*)

Optical properties		Dumortierite	Enstatite	Epidote	Fluorite	Garnet
Plane polarized light	Pleochroism/colour	Pleochroic, colourless, red deep blue	Nonpleochroic, colourless	Pleochroic, pale greenish yellow to yellowish green (lemon yellow)	Nonpleochroic, colourless	Nonpleochroic, colourless, pink, yellow
	Refractive index	Moderate	High	High	Low	High
	Shape and habit	Prismatic, striations parallel to prism edge	Prismatic	Anhedral, angular to subrounded	Euhedral or anhedral	Angular to subrounded, often with conchoidal fracture
	Cleavage	Imperfect	2 sets perfect	Imperfect	Imperfect	Irregular fractures
	Alteration	Uncommon	Antigorite	Uncommon	Uncommon	Uncommon
	Inclusion	Uncommon	Common	Uncommon	Common	Common
Crossed nicol	Optic character	Biaxial –	Biaxial +	Biaxial –	Isotropic	Isotropic
	Interference colour	First and second orders	First order white to gray	High order, same as in ordinary light, anomalous blue or brownish	—	—
	Birefringence	Low to moderate	Low	High	—	—
	Extinction	—	Parallel	—	—	—
	Twinning	Uncommon	Rare	Common	—	—
	Sign of elongation	Negative	Positive	Positive/negative	—	—
	Optic axial angle	30°–40°	58°–80°	69°–89°	—	—

(Contd...)

TABLE 14.4: Optical properties of some common transparent heavy minerals (*Contd...*)

Optical properties		Glaucophane	Hornblende	Hypersthene
Plane polarized light	Pleochroism/colour	Pleochroic, colourless to violet blue and dark blue	Pleochroic, ccommon hb.—brownish green to greenish brown to black; metamorphic hb. – bluish green; basaltic hb.—reddish brown	Pleochroic, reddish pink to yellowish green
	Refractive index	High	High	High
	Shape and habit	Prismatic	Elongate prismatic	Prismatic
	Cleavage	2 sets perfect	2 sets perfect	2 sets
	Alteration	Uncommon	Chlorite, epidote	Serpentine, cloudy due to alteration
	Inclusion	Uncommon	Uncommon	Uncommon
Crossed nicol	Optic character	Biaxial –	Biaxial –	Biaxial –
	Interference colour	Second order violet	Second order colours	First order gray
	Birefringence	Moderate	Moderate	Low
	Extinction	Oblique, 4°–6°	Oblique, 15°–25°	Parallel
	Twinning	Absent	Common	Absent
	Sign of elongation	Positive	Positive	Positive
	Optic axial angle	0°–68°	52°–85°	63°–90°
	Twinning	Uncommon	Rare	Common —
	Sign of elongation	Negative	Positive	Positive/negative —
	Optic axial angle	30°–40°	58°–80°	69°–89° —

(*Contd...*)

TABLE 14.4: Optical properties of some common transparent heavy minerals (*Contd...*)

Optical properties		Kyanite	Leucoxene	Monazite
Plane polarized light	Pleochroism/colour	Nonpleochroic, colourless, rarely pale blue	Nonpleochroic, dead white	Weakly pleochroic, colourless or yellow to pale greenish yellow
	Refractive index	High	Low to moderate	High
	Shape and habit	Anhedral, rectangular	Anhedral, often associated with ilmenite	Anhedral, equidimensional, rounded (egg shaped)
	Cleavage	Conspicuous cross cleavage	Absent	Absent
	Alteration	Common (to mica)	Uncommon	Uncommon
	Inclusion	Common	Uncommon	Uncommon
Crossed nicol	Optic character	Biaxial –	—	Biaxial +
	Interference colour	Good cleavage in three directions show step like areas of bright first and second order colours	First order white (same colour as in ordinary light)	Third to fourth order (many grains show the same colour as in ordinary light)
	Birefringence	Low – moderate	—	High
	Extinction	7°–30°	—	2°–10°
	Twinning	Common	—	Absent
	Sign of elongation	Positive/negative	—	—
	Optic axial angle	82°–83°	—	11°–14°

(Contd...)

TABLE 14.4: Optical properties of some common transparent heavy minerals (Contd...)

	Optical properties	Muscovite	Rutile	Siderite	Sillimanite
Plane polarized light	Pleochroism/colour	Nonpleochroic, colourless	Nonpleochroic to feebly pleochroic deep red or yellow	Nonpleochroic, brown, yellow	Nonpleochroic, colourless
	Refractive index	Low	Very high	High	High
	Shape and habit	Equant, rests on basal cleavage	Prismatic, rounded, striated, thick boarders around the grain	Subangular, rhombohedral, spherulitic, colloform	Short prismatic, fibrous, striations parallel to length
	Cleavage	1 set perfect	1 set	2 sets rhombohedral	1 set
	Alteration	Clay minerals	Uncommon	Clay minerals	—
	Inclusion	—	Abundant	Uncommon	Uncommon
Crossed nicol	Optic character	Biaxial –	Uniaxial +	Biaxial –	Biaxial +
	Interference colour	First order gray, black on basal cleavage	Deep red or yellow (same as body colour)	First order gray, white	Second to third order rainbow colours
	Birefringence	Very low when rest on basal cleavage	Very high	Low	Moderate to high
	Extinction	Parallel	Parallel	Parallel to 5°	Parallel
	Twinning	Uncommon	Knee shaped	Simple twin	Absent
	Sign of elongation	—	Positive	—	Positive
	Optic axial angle	29°–42°	—	0°–12°	20°–25°
	Sign of elongation	Positive/negative	—	—	Positive
	Optic axial angle	82°–83°	—	11°–14°	—

(Contd...)

TABLE 14.4: Optical properties of some common transparent heavy minerals (*Contd...*)

	Optical properties	Sphene	Spinel	Staurolite	Topaz
Plane polarized light	Pleochroism / colour	Feebly pleochroic, colourless to pale yellow or brownish	Nonpleochroic, grass green or coffee brown	Pleochroic, colourless to golden yellow, (occasionally straw yellow)	Nonpleochroic, colourless
	Refractive index	Very high	High	High	High
	Shape and habit	Euhedral (diamond-shaped) to anhedral, conchoidal	Subhedral to well rounded	Prismatic, anhedral to angular with hackly to subconchoidal fracture	Columnar
	Cleavage	Absent	Imperfect	Imperfect	1 set perfect
	Alteration	Uncommon	Uncommon	Uncommon	Uncommon
	Inclusion	Uncommon	Uncommon	Plenty	Fluid
Crossed nicol	Optic character	Biaxial +	Isotropic	Biaxial –	Biaxial +
	Interference colour yellow, red	Same as in ordinary light First order yellow	—	First order	
	Birefringence	Very high	—	Low	Low
	Extinction	Incomplete, bluish at extinction	—	Parallel	Parallel to symmetrical
	Twinning	Uncommon	—	Rare	Absent
	Sign of elongation	—	—	Positive	Negative
	Optic axial angle	27°–35°	—	80°–88°	48°–66°

(*Contd...*)

TABLE 14.4: Optical properties of some common transparent heavy minerals (*Contd…*)

Optical properties		Tourmaline	Tremolite	Zircon	Zoisite
Plane polarized light	Pleochroism/colour	Strongly pleochroic, yellow to brown, pink to dark; indicolite—indigo blue to black	Feebly pleochroic, colourless to pale green or brown	Nonpleochroic to feebly pleochroic, colourless, pink, reddish or mauve	Feebly pleochroic, colourless to pink or green
	Refractive index	High	High	Very high	High
	Shape and habit	Prismatic, occasionally well rounded or oval, often with striations	Prismatic, fibrous	Euhedral, prismatic or rounded with dark boarder	Euhedral, prismatic, anhedral
	Cleavage	Imperfect	2 sets perfect	Absent	1 set perfect
	Alteration	Uncommon	Talc	Uncommon	Uncommon
	Inclusion	Common	Uncommon	Plenty	Uncommon
Crossed nicol	Optic character	Uniaxial –	Biaxial –	Uniaxial +	Biaxial +
	Interference colour	Second and third order colours	Second order	Fourth order	Abnormal ultra blue
	Birefringence	High	High	Very high	Low to moderate
	Extinction	Parallel	Inclined, 10°–20°	Parallel	Incomplete
	Twinning	Absent	Uncommon	Absent	Uncommon
	Sign of elongation	Negative	Positive	Positive	Positive
	Optic axial angle	—	80°–85°	—	20°–60°

(*Contd…*)

counting or may be counted twice (H), though it is a single crystal. To overcome these difficulties, it is suggested that the *field count* method should be followed for determination of heavy mineral frequency. In this method, all the heavy minerals seen in one field of view are taken into consideration and counted. Then the slide is moved till adjacent new field of view comes under the microscope. The process is repeated till the last field of view at other end of the slide is reached. Some authors recommend the *ribbon count* method in which the minerals lying within a rectangular strip of area on both side of the cross wires are taken into consideration.

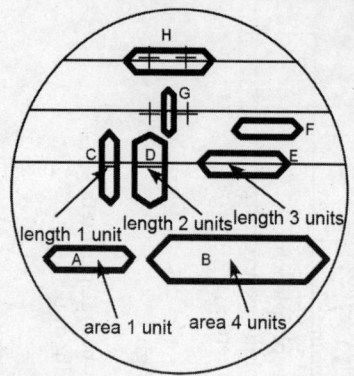

Fig. 14.4: Counting of heavy minerals

14.5 GRAPHICAL PRESENTATION OF HEAVY MINERAL FREQUENCY

In many instances the frequencies of heavy minerals vary within wide limits, as a result of which, pictorial presentation in form of pi- and bar-diagrams in arithmetic scale create problem for easy visualization. It is recommended that the heavy mineral frequencies should be presented in log-scale.

Example: On the average, the heavy mineral assemblage of the Barakar sandstones of the Talchir Gondwana basin is constituted of opaque, 23.49%; leucoxene, 3.46%; garnet, 61.65%; zircon, 3.42%; tourmaline, 0.29%; rutile, 0.31%; biotite, 1.22%; chlorite, 0.31%; pyroxene, 0.62%; hornblende, 1.87%; hypersthene, 0.44%; sillimanite, 0.35%; staurolite, 0.11%; spinel, 0.03%; apatite, 0.05%; epidote, 0.03%; sphene, 0.23% and siderite, 2.12%.

Since the frequency of heavy mineral species varies from 0.03% to 61.65%, the frequency distribution cannot be shown by ordinary arithmetic-ordinate scale. The frequency distribution of heavy minerals of the Barakar sandstone of the Talchir Gondwana basin is shown in Fig. 14.5, in logarithm-ordinate scale. The observed heavy mineral suite suggests their derivation from a composite source consisting of pegmatite, acid and basic igneous rocks as well as low- to high-grade metamorphic rocks.

Since the opaque mineral species cannot be identified in transmitted polarized light, many authors advocate counting of the opaque and transparent fractions separately. Even, the secondary and authigenic minerals are to be excluded as they do not provide any clue regarding the source rock lithology. Thus, the heavy mineral suite of the Barakar formation consists of opaque, 23.49%; leucoxene (secondary), 3.46%; siderite (derived from the basin), 2.12% and transparent heavy minerals, 70.93%. By recalculation the list becomes garnet, 86.92%; zircon, 4.82%; tourmaline, 0.41%; rutile, 0.44%; biotite, 1.72%; chlorite, 0.44%; pyroxene, 0.87%; hornblende, 2.64%; hypersthene, 0.62%; sillimanite, 0.49%; staurolite, 0.16%; spinel, 0.04%; apatite, 0.07%; epidote, 0.04% and sphene, 0.32%. Zircon, tourmaline and rutile are the most stable heavy minerals. The sum of these three minerals is known as ZTR index. It is defined as the sum of zircon, tourmaline and rutile among the nonmicaceous transparent heavy minerals. So, minerals like biotite and chlorite are to be excluded. The ZTR index is regarded as an indicator of heavy mineral maturity. By recalculation, the amount of zircon, tourmaline and rutile become 4.93%, 0.42% and 0.45% respectively. So the ZTR index for the Barakar sandstone of Talchir Gondwana basin is 5.8.

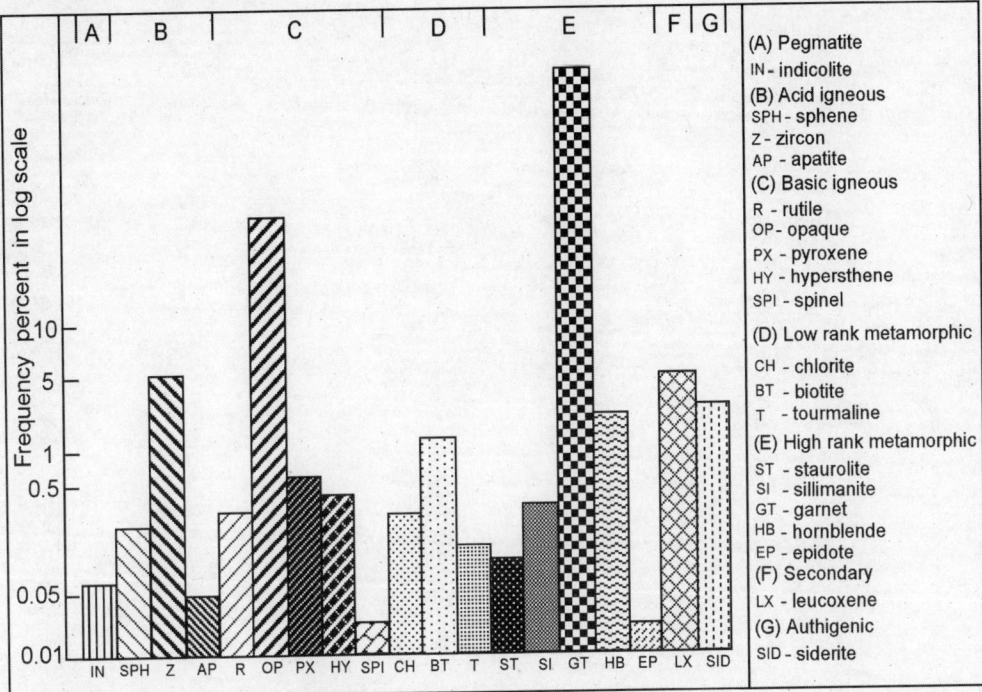

Fig. 14.5: Graphic representation of heavy minerals of the Barakar formation, Talchir Gondwana basin

14.6 SIGNIFICANCE OF HEAVY MINERAL STUDY

The heavy mineral study helps in

   i. Deciphering the provenance lithology

  ii. Study of sediment dispersal pattern

 iii. Identification of unconformities

 iv. Palaeogeographic reconstruction

  v. Stratigraphic correlation

 vi. Marine regression characteristics

vii. Evaluation of diagenetic history

viii. Discovery of placer minerals.

Bibliography

1. Best, MG. (1986) Igneous and metamorphic petrology. CBS Publishers and Distributors, New Delhi.

2. Bose, MK. (1997) Igneous petrology. The World Press Pvt. Ltd., Kolkata.

3. Collinson, JD and Thompson, DB (1989) Sedimentary structures. CBS Publishers and Distributors, New Delhi.

4. Davis, JC. (2002) Statistics and data analysis in geology. John Wiley and Sons, New York.

5. Gokhale, NW. (1998) Fundamentals of Sedimentary rocks. CBS Publishers and Distributors, New Delhi.

6. Herpum, JR. (1963) Petrographic classification of granitic rocks by partial chemical analysis. Tanganyika Geol. Surv. Rep, v.10, pp.80–86.

7. Hota, RN. (1999) Subsurface geology of the Barakar Formation, Talchir Gondwana basin, Orissa, India. Unpublished PhD thesis, Utkal University, p.237.

8. Hota, RN. (2007) Palaeocurrent pattern and its tectonic implication during Talchir and Damuda sedimentation in Talchir Gondwana basin, Orissa. Gondwana Geological Magazine, v. 22 (No.1), pp.1–10.

9. Hota, R.N., Maejima, W. and Mishra, B. (2007) River metamorphosis during Damuda sedimentation: a case study from the Talchir Gondwana basin, Orissa. Journal of Geological Society of India, v.69, pp.1351–1360.

10. Hutchison. CS. (1974) Laboratory handbook of petrographic techniques. John Wiley and Sons, New York.

11. Le Bas, MJ and Streckeisen, AL (1991) The IUGS systematics of 3 igneous rocks. Jour. Geol. Soc. London, v.148, pp.825–833.

12. Lindholm, R. (1987) A practical approach to sedimentology. Allen and Unwin, London.

13. McDonald, G. A. and Katsura, T. (1964) Chemical composition of Hawaiian lavas. Jour. Petrol, v.5, pp.82–133.

14. Mason, R. (1990) Petrology of the Metamorphic Rocks. Unwin Hyman, London.

15. Miyashiro, A. (1994) Metamorphic petrology. UCL Press Ltd, London.

16. Mullen, E. (1983) $MnO - TiO_2 - P_2O_5$—a major element discriminant for basaltic rocks of ocean environments and its implications for petrogenesis. Earth and Planet. Sci. Let. v.65, pp.41–58.

17. Nichols, Gary (1999) Sedimentology and stratigraphy. Blackwell Science Ltd., London.

18. O'Connor, JJ. (1965) A classification for the quartz rich igneous rocks based on feldspar ratios. US Geol. Prof. paper, 525p.

19. Pettijohn, FJ. (1963) Chemical composition of sandstone—excluding carbonate and volcanic sand. U.S. Geol. Surv. Prof. Paper, v.440-S, 21p.

20. Pettijohn, FJ. (1984) Sedimentary rocks. CBS Publishers and Distributors, New Delhi.

21. Prothero, RP and Schwab, F. (1996) Sedimentary Geology. WH freeman and Co., New York.

22. Sen, AK. (1995) Laboratory manual of Geology. Modern Book Agency Pvt., Ltd, Kolkata.

23. Sengupta, SM. (2007) Introduction to sedimentology. CBS Publishers and Distributors, New Delhi.

24. Sukhtankar, RK. (2004) Applied Sedimentology. CBS Publishers and Distributors, New Delhi.

25. Swan, ARH and Sandilands, M. (1995) Introduction to geological data analysis. Blackwell Science Ltd. London.

26. Turner, FJ. and Verhoogen, J. (1987) Igneous and metamorphic petrology. CBS Publishers and Distributors, New Delhi.

27. Tyrrell, GW. (1977) The Principles of Petrology. Asia Publishing House, New Delhi.

28. Walker, KR, Joplin, GA, Lovering, JF and Green, R. (1960) Metamorphic and metasomatic convergence of basic igneous rocks and lime magnesium sediments of the Precambrian of Northwest Queenland. Jour. Geol. Soc. Australia, v.6, pp.149–178.

29. Whitten, DG. A and Brooks, JRV. (1972) The Penguin dictionary of Geology. Penguin Books Ltd., Harmondsworth.

30. Winkler, HGF. (1976) Petrogenesis of Metamorphic rocks. Narosa Publishing House, New Delhi.

Index

ACF diagram 59
Acid-charnockite 56
Actinolite 151
Adamellite 13
AFM diagram 62
AKF diagram 61
Amphibolite 56
Amygdaloidal structure 4
Analysis of grain size data 112
Anatase 151
Andalusite 151
Andesite 13
Anorthosite 13
Apatite 151
Aplite 13
Arenicolites 90, 97
Arenite 102
Armored mud balls 83
Asterosoma 90, 97
Augen-gneiss 56
Augite 151

Barite 152
Basalt 13
Basic granulite 56
Bedding 76
Bergaueria 90, 97
BHJ 56
Bifungites 90, 97

Biogenic structure 86
Bio-micrudite 107
Bio-sparudite 107
Biotite 150
Blocky structure 3
Borings 87
Brookite 150
Burrows 87

Calc-schist 56
Calc-silicate 56
Carbonatite 13
Cassiterite 152
Cataclastic
    structure 55
    texture 54
Cement 74
Charnockite 13, 56
Chemical structure 83
Chlorite 153
Chlorite-schist 56
Chloritoid 153
Chondrites 91, 97
Classification of
    carbonate rocks (limestone) 103
    conglomerates and breccias 101
    Friedman and Sander 100
    mud rocks 102
    Pettijohn 100

sandstones 102
sedimentary aggregates 104
Clastic texture 73
Coal 107
Columnar structure 5
Concretion 84
Cone-in-cone 84
Coprolite 87
Corrosion surface 85
Corundum 151
Counting of heavy minerals 147
Cross
    bedding 78
    lamination 78
Cruziana 91, 97
Cryptocrystalline 6
Crystal mould 86
Crystallinity, igneous rock 6
Crystalloblastic texture 54
Cylindrichnus 91, 97

Dacite 13
Diabase 13
Diagenetic features 74–75
Diopside 151
Diorite 13
Diplichnites 91, 97
Diplocraterion 91, 97
Directive texture 8
Discrimination diagram 47–50

Dolerite 14
Dumortierite 154
Dunite 14

Eclogite 14, 56
Enstatite 152
Epidote 152
Essexite 14

Fabric 72
Faecal pellet 87
Felsite 14
Flagstone 106
Flaser bedding 79
Flow structure 4
Fluorite 154
Flute cast 81
Fracture form 9

Gabbro 14
Garnet 154
Geode 84
Glaucophane 155
Gneiss 57
Gneissose structure 55
Gondite 57
Graded bedding 81
Granite 14
   gneiss 57
Granoblastic texture 54
Granodiorite 14
Granophyre 14
Granularity, igneous rock 6
Granulite 57
Granulose structure 54
Graphical presentation of
   heavy mineral 160–161
Greenstone 57
Groove cast 81

Harker diagram 44
Heavy mineral
   separation 146–147
Helminthoida 92, 97

Herring-bone cross-
   stratification 80
Histogram 115
Hornblende 155
Hornblende-gneiss 57
Hornblende-schist 57
Hornblendite 15
Hornfels 57
Hornfelsic texture 54
Hummocky cross-
   stratification 80
Hypersthene 155
   granulite 57

Ichnofossil 88
Intergranular texture 8
Intergrowth texture 8
Intersertal textures 8
Intraclasts 103

Joint igneous rock 4

Khondalite 57
Kimberlite 15
Kurtosis 113, 114
Kyanite 154

Lamination 76
Lamprophyre 15
Laterite 107
Lava 3
Lenticular bedding 79
Lepidoblastic texture 54
Leptynite 57
Leucoxene 156
Limestone 107
Load cast 82

Maculose structure 55
Magma 3
Magnetite quartzite 58
Marble 58, 107
Marl 107
Matrix 74

Mechanical structure 76
Median 113
Meta conglomerate 58
Micrite 103, 107
Microcrystalline 6
Migmatite 58
Modal analysis 40
   of igneous rock 41–42
   of sedimentary rock 42–43
Monazite 156
Monocraterion 92, 98
Monzonite 15
Mounting of heavy minerals
Mud crack cast 82
Mudstone 107
Muscovite 155
Mylonite 58

Nematoblastic texture 54
Nepheline syenite 15
Nereites 92, 98
Niggli values 23–24
Nodule 83
Nonclastic texture 73
Norite 15
Norm 17
Normative classification 22

Obsidian 15
Oligomictic breccia 106
Oligomictic conglomerate 106
Oolicast 85
Oosparite 107
Ophiomorpha 92, 98
Orbicular structure 9
Organic texture 73

Packing 72–73
Palaeocurrent 124–139
Palaeohydrology 142–144
Palaeophycus 93, 98
Paleodictyon 93, 98
Pegmatite 15